SPEND

Governments in some democracies target economic policies, like industrial subsidies, to small groups at the expense of many. Why do some governments redistribute more narrowly than others? Their willingness to selectively target economic benefits, like subsidies to businesses, depends on the way politicians are elected and the geographic distribution of economic activities. Based on interviews with government ministers and bureaucrats, as well as parliamentary records, industry publications, local media coverage, and new quantitative data, *Spending to Win: Political Institutions, Economic Geography, and Government Subsidies* demonstrates that government policy-making can be explained by the combination of electoral institutions and economic geography. Specifically, it shows how institutions interact with economic geography to influence countries' economic policies and international economic relations. Identical institutions have wide-ranging policy effects depending on the context in which they operate. No single institution is a panacea for the pressing issues facing many democracies today, such as income inequality, international economic conflict, or representation.

Stephanie J. Rickard is an Associate Professor at the London School of Economics.

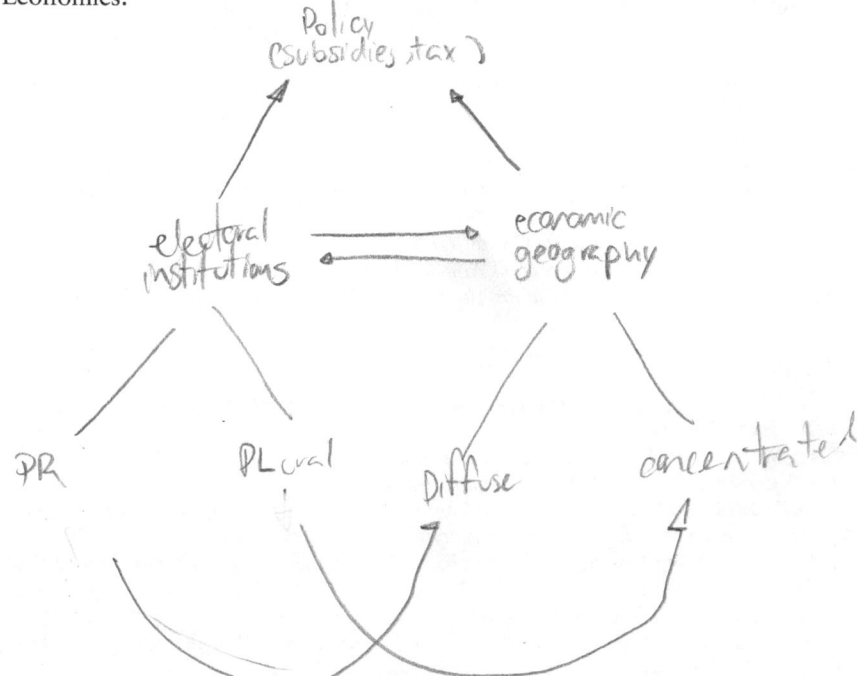

POLITICAL ECONOMY OF INSTITUTIONS AND DECISIONS

Series Editors

Jeffry Frieden, *Harvard University*
John Patty, *University of Chicago*
Elizabeth Maggie Penn, *University of Chicago*

Founding Editors

James E. Alt, *Harvard University*
Douglass C. North, *Washington University of St. Louis*

Other books in the series

Alberto Alesina and Howard Rosenthal, Partisan Politics, *Divided Government and the Economy*
Lee J. Alston, Thrainn Eggertsson and Douglass C. North, eds., *Empirical Studies in Institutional Change*
Lee J. Alston and Joseph P. Ferrie, *Southern Paternalism and the Rise of the American Welfare State: Economics, Politics, and Institutions, 1865–1965*
James E. Alt and Kenneth Shepsle, eds., *Perspectives on Positive Political Economy*
Josephine T. Andrews, *When Majorities Fail: The Russian Parliament, 1990–1993*
Jeffrey S. Banks and Eric A. Hanushek, eds., *Modern Political Economy: Old Topics, New Directions*
Yoram Barzel, *Economic Analysis of Property Rights, 2nd edition*
Yoram Barzel, *A Theory of the State: Economic Rights, Legal Rights, and the Scope of the State*
Robert Bates, *Beyond the Miracle of the Market: The Political Economy of Agrarian Development in Kenya*
Jenna Bednar, *The Robust Federation: Principles of Design*
Charles M. Cameron, *Veto Bargaining: Presidents and the Politics of Negative Power*
Kelly H. Chang, *Appointing Central Bankers: The Politics of Monetary Policy in the United States and the European Monetary Union*
Peter Cowhey and Mathew McCubbins, eds., *Structure and Policy in Japan and the United States: An Institutionalist Approach*
Gary W. Cox, *The Efficient Secret: The Cabinet and the Development of Political Parties in Victorian England*
Gary W. Cox, *Making Votes Count: Strategic Coordination in the World's Electoral System*
Gary W. Cox, *Marketing Sovereign Promises: Monopoly Brokerage and the Growth of the English State*
Gary W. Cox and Jonathan N. Katz, *Elbridge Gerry's Salamander: The Electoral Consequences of the Reapportionment Revolution*
Tine De Moore, *The Dilemma of the Commoners: Understanding the Use of Common-Pool Resources in Long-Term Perspective*
Adam Dean, *From Conflict to Coalition: Profit-Sharing Institutions and the Political Economy of Trade*
Mark Dincecco, *Political Transformations and Public Finances: Europe, 1650–1913*
Mark Dincecco and Massimiliano Gaetano Onorato, *From Warfare to Wealth: The Military Origins of Urban Prosperity in Europe*

(continued after index)

SPENDING TO WIN

*Political Institutions, Economic Geography,
and Government Subsidies*

STEPHANIE J. RICKARD
London School of Economics and Political Science

CAMBRIDGE
UNIVERSITY PRESS

University Printing House, Cambridge CB2 8BS, United Kingdom

One Liberty Plaza, 20th Floor, New York, NY 10006, USA

477 Williamstown Road, Port Melbourne, VIC 3207, Australia

314-321, 3rd Floor, Plot 3, Splendor Forum, Jasola District Centre, New Delhi - 110025, India

79 Anson Road, #06-04/06, Singapore 079906

Cambridge University Press is part of the University of Cambridge.

It furthers the University's mission by disseminating knowledge in the pursuit of education, learning and research at the highest international levels of excellence.

www.cambridge.org
Information on this title: www.cambridge.org/9781108432030
DOI: 10.1017/9781108381475

© Stephanie J. Rickard 2018

This publication is in copyright. Subject to statutory exception and to the provisions of relevant collective licensing agreements, no reproduction of any part may take place without the written permission of Cambridge University Press.

First published 2018
First paperback edition 2020

A catalogue record for this publication is available from the British Library

ISBN 978-1-108-42232-1 Hardback
ISBN 978-1-108-43203-0 Paperback

Cambridge University Press has no responsibility for the persistence or accuracy of URLs for external or third-party internet websites referred to in this publication, and does not guarantee that any content on such websites is, or will remain, accurate or appropriate.

For Shane

Contents

List of Figures		*page* viii
List of Tables		ix
Acknowledgments		x
1	Who Gets What and Why? The Politics of Particularistic Economic Policies	1
2	The Uneven Geographic Dispersion of Economic Activity	27
3	How Institutions and Geography Work Together to Shape Policy	39
4	Explaining Government Spending on Industrial Subsidies	64
5	The Power of Producers: Successful Demands for State Aid	97
6	Why Institutional Differences among Proportional Representation Systems Matter	134
7	The Policy Effects of Electoral Competitiveness in Closed-List PR	170
8	Conclusion and Implications	199
References		216
Index		236

Figures

4.1	Marginal effect of proportional representation (PR) on subsidy budget shares	page 87
4.2	Marginal effect of disproportionality on subsidy budget shares	91
5.1	Predicted number of non-EU compliant subsidies	103
5.2	Total national government financial aid to the wine industry, in millions of euros	109
6.1	Marginal effect of open lists on subsidy budget shares in PR countries	150
6.2	Variation in the geographic concentration of economic sector employment in Norway	155
6.3	Marginal effect of mean district magnitude on subsidy budget shares in open-list PR	165
7.1	Average subsidy amount per manufacturing employee, 2005–2012	174
7.2	Electoral disproportionality over time in Norway	178
7.3	Largest government party's vote margin and subsidies per employee, by district	195

Tables

3.1	Illustration of empirical expectations	page 51
4.1	Effects of institutions and geography on subsidy budget shares	84
4.2	Effects of various features of countries' electoral systems on subsidy budget shares	89
4.3	Second stage results of the effects of PR on subsidy budget shares	94
5.1	Effects of PR on non-EU compliant subsidies	102
5.2	Parliamentary questions about the French Cognac subsidy	115
5.3	Electoral competitiveness in Charentes	119
5.4	Austrian farm-gate wine sales as a percentage of total sales	121
6.1	Effect of open-party lists on subsidy budget shares	148
6.2	Second-stage results of the effect of open-party lists on subsidy budget shares	153
6.3	Effect of geographic concentration on sector-specific subsidies in a closed-list PR system	160
6.4	Effect of mean district magnitude on subsidies in open-list PR	163
6.5	Effect of mean district magnitude on subsidies in closed-list PR	167
7.1	Explaining the variation in manufacturing subsidies per employee between electoral districts	194

Acknowledgments

This book is the culmination of a long journey. Along the way, I have amassed so many debts that I cannot possibly acknowledge all of them here. The journey began at the University of Rochester, where Jim Johnson, Ron Jones, and Randy Stone generously took the time and interest to set me on this path. At UCSD, I was privileged to work with the very best mentors in the business: Lawrence Broz, Steph Haggard, and David Lake. I owe a huge debt to them all.

The main idea in this book first came to me while I was visiting the Institute for International Integration Studies (IIIS) at Trinity College Dublin. I am grateful to Ken Benoit, Philip Lane, and Kevin O'Rourke for inviting me to visit and their gracious hospitality during my stay.

Thank you to Rob Franzese and Ken Scheve for encouraging me to write this book. Without their initial encouragement, this book would not exist. Thank you to David Singer who nudged me along the way and generously provided me with a sanctuary at MIT to write while I was on sabbatical. David read and commented on so many drafts that I will be forever in his debt.

I am grateful to everyone who read and commented on sections of the manuscript including Lucy Barnes, Sarah Brooks, Lawrence Broz, John Carey, José Cheibub, Alexandra Cirone, Raphael Cunha, Christina Davis, David Doyle, Jeff Frieden, Miriam Golden, Sara Hagemann, Simon Hix, Sara Hobolt, Mark Kayser, Yuch Kono, Marcus Kurtz, Karen Long Jusko, David Lake, David Leblang, Helen Milner, Shane Martin, Layna Mosley, Wolfgang Müller, Bjørn Erik Rasch, Ron Rogowski, Jonathan Rodden, Petra Schleiter, David Singer, Kaare Strøm, David Stasavage, Guido Tabellini, Daniel Verdier, and Rachel Wellhausen. A special thanks to Robert Elgie who not only read and commented on my French case but also helped me acquire the French legislative documents. Thank you also to Marius Brülhart and Vincent Pons for generously sharing their data with me.

Many friends and colleagues offered support and advice along the way for which I am grateful, including Kristin Bakke, Teri Caraway, Mark Copelovitch, Manfred Elsig, Judy Goldstein, Nate Jensen, Soo Yeon Kim, Quan Li, Ed Mansfield, Megumi Naoi, Dan Nielson, Sonal Pandya, Pablo Pinto, Dennis Quinn, Nita Rudra, Ken Shadlen and Erik Wibbels. Thank you.

I was privileged to have the opportunity to present parts of this project and receive helpful feedback from seminar participants at Essex University, Duke University, University of Illinois, McGill University, Ohio State University, Oxford University, Princeton University, University College London, University of Kent, the World Trade Institute and Yale University.

This research was supported by funding from the Research Council of Norway (FriSam Project No. 222442). I am thankful to Bjørn Erik Rasch, the Principal Investigator, who kindly hosted me at the University of Oslo while I was on sabbatical and read and commented on my Norwegian case.

I am also thankful for the staff at Innovation Norway. Everyone I spoke with at Innovation Norway was generous with their time and admirably frank about how government-funded subsidies work in practice. I am especially grateful to Sigrid Gåseidnes who went out of her way to help me.

Finally, I would never have finished this book without the unending support of Shane Martin. Not only did he take on an unfair share of work to give me the time I needed to write, he also provided invaluable intellectual guidance. People familiar with Shane's research will see evidence of his influence on my thinking about politics in this book. Shane read and commented on more drafts that one would think humanly possible and his boundless positivity and encouragement kept me going. I am very lucky to have such an amazing partner in life. Thank you.

I

Who Gets What and Why?

The Politics of Particularistic Economic Policies

Democratic institutions ostensibly serve the common good. Yet democratically elected leaders face diverse incentives. Politicians must balance the public's welfare with demands from interest groups that run counter to the common good. Nowhere is this balancing act more apparent than in the area of economic policy. Governments' economic policies often redistribute resources between groups. Governments collect money from taxpayers and then spend the tax revenues on various programs. Governments may fund programs that support broad groups of citizens, such as health care or education. Alternatively, governments may use their fiscal resources to privilege small, select groups of citizens via programs like subsidies for business.[1] Subsidies typically provide economic benefits selectively to small groups and accordingly can be described as "particularistic" economic policies. Although particularistic economic policies often entail costs for many citizens, including taxpayers and consumers, they nonetheless emerge in democratic contexts.

Although the political motivations behind particularistic economic policies have been studied extensively,[2] the variation in such policies between countries is less well understood. Leaders in some democratic countries enact more particularistic economic policies than others and as a result, particularistic economic policies vary in both frequency and magnitude among democracies. In France and Australia, for example, leaders habitually provide narrowly targeted financial assistance to select businesses. Similarly, governments in the United Kingdom subsidized individual firms during the 1960s and 1970s, including state-

---

[1] Governments can also privilege select groups by exempting them from paying taxes.
[2] Producers' demands often prevail because they are fewer in number and consequently can organize more easily than taxpayers and consumers (e.g. Olson 1965, Alt and Gilligan 1994). Producers also have more at stake. Government subsidies can mean the difference between bankruptcy and profit. However, for taxpayers, the cost of any given subsidy program is negligible. Taxpayers consequently have few incentives to oppose subsidies.

owned companies like British Steel and British Airways (Sharp, Shepherd, and Marsden 1987). In contrast, during the same period, the West German government refused to provide subsidies to individual firms (Schatz and Wolter 1987). The government focused instead on building a comprehensive framework of policies that would benefit large numbers of citizens called the *Soziale Marktwirtschaft* (Sharp et al. 1987). Today, governments in some countries continue in the same tradition by providing general assistance to broad groups, as in Finland (Verdier, 1995: 4).

The diversity in democracies' economic policies reflects the varied responsiveness of politicians to different interests. In France, for example, most politicians believe it is their duty above all else to represent the citizens living in their geographically-defined electoral district.[3] French politicians consequently work hard to secure economic benefits for their constituents. As one member of the French parliament (MP) colorfully put it, "[a]n MP is a gardener. He has a big garden – his constituency – and he has to go to Paris in order to get fertiliser."[4] Particularistic economic policies can provide such fertilizer.

In contrast, leaders in some countries strive to represent larger groups of citizens. In Sweden, for example, the government refused to bail out the ailing automotive industry following the 2008 global financial crisis. The Prime Minister said he would not put "taxpayer money intended for healthcare or education into owning car companies" (Ward 2009). The German government similarly resisted demands for industrial subsides in the wake of the 2008 crisis. Defending this decision, the German Minister for Economics and Technology said that privileging certain industries went against "all successful principles of our economic policy."[5] Governments' varied approaches to economic policy highlight the puzzle motivating my book: why do governments provide more particularistic economic policies in some democracies than others?

The goal of this book is to understand economic policy. Specifically, I seek to explain the variation in economic support provided by democratically elected governments to firms, industries, and sectors, such as manufacturing. Understanding why governments do more to assist such groups in some countries is important. Democratic theorists have long worried about the power of special interests. Groups that pursue economic rents for themselves at the expense of others are of

---

[3] In a survey conducted by Brouard et al. (2013), 41.2 percent of sitting French MPs said this.
[4] Quoted in Brouard et al. (2013: 146). This quote came from a member of the centre-right political party Union for a Popular Movement (*Union pour un mouvement populaire*/UMP).
[5] *The Economist* November 1, 2008: 62.

particular concern. Groups seeking government subsidies "not only pervert the meaning of democratic accountability but also create deadweight losses and distort economic incentives" (Cox and McCubbins, 2001: 48).[6] The money governments spend on subsidies is money no longer available for other programs, such as education or health care. And governments facing tight budgets often cut programs, such as social welfare, in order to fund increased spending on subsidies (Rickard 2012b). Subsidies consequently have serious implications for the regressivity of government spending. It is therefore important to understand why leaders in some democracies spend relatively more on particularistic economic policies, like subsidies.

I argue that economic policy outcomes depend on the way politicians are elected and the distribution of economic activities in space. Economic policy cannot be explained by political institutions alone, contrary to conventional wisdom. Economic geography – that is, the geographic distribution of economic activities – must also be considered to understand governments' economic policy decisions. My argument stands in contrast to "pure" institutional arguments that identify political institutions as the key determinant of countries' economic policies (e.g. Persson and Tabellini 2003). This book's core thesis is that economic policies result from the interactive effects of economic geography and political institutions, specifically the institutions governing democratic elections. Electoral institutions determine the optimal (re)election strategy for politicians and political parties competing in democratic elections. Economic geography determines which economic policies best accomplish the institutionally generated electoral strategy. In the following section, I briefly outline the contours of my argument, which I develop more fully in Chapter 3.

ARGUMENT IN BRIEF

Elections aggregate voters' preferences. But not all elections work the same way. Different rules govern election contests in different countries. The rules governing elections, often referred to as electoral institutions or electoral systems, determine how elections work and ultimately how elections aggregate voters' preferences. To understand the effects of electoral institutions, it is important to know what voters want. Equally important, however, is knowing where voters with shared preferences live. Voters with shared policy preferences may live close to one another in relatively small, geographically concentrated areas. But voters with shared policy preferences may alternatively reside throughout the

---

[6] See also Stigler (1971) and Becker (1985).

country. Knowing where voters with shared policy preferences live is vital to understanding how electoral institutions shape policy outcomes because different electoral institutions provide dissimilar incentives for politicians to respond to groups with different geographic characteristics.

Voters' economic policy preferences depend, in part, on their economic security. Most voters work to earn a living and as a result, their personal economic prosperity is closely tied to the economic fortunes of their employer.[7] People want their employer to be economically successful because successful industries hire and retain more employees, typically offering more generous wages and compensation packages (Aghion et al. 2011, Criscuolo et al. 2012, Stöllinger and Holzner 2016). Industries' ability to pay generous wages and provide secure employment opportunities often depends on governments' economic policies, including, for example, subsidies.

People employed in a given industry share a common interest in the economic performance of the industry and government policies that promote its performance.[8] This shared interest is "narrow" because most industries typically employ only a small fraction of a country's total population. The US steel industry, for example, employs only 0.3 percent of the US population. The steel industry therefore constitutes a "narrow" or "special" interest, as defined here.

Narrow interests can be more or less geographically concentrated depending on the geographic patterns of employment. Although industries today have fewer constraints on where they locate and employees tend to be more geographically mobile, strong patterns of geographic concentration persist at both a national and regional level in many economies (Krugman 1991, OECD 2008, Autor, Dorn, and Hanson 2013). But not all industries are equally concentrated (Autor et al. 2013). While employees in the US steel industry are primarily located in just three of the fifty US states, the tourism industry, in contrast, employs people across the entire country. As these illustrative examples make clear, different industries have varied geographic patterns of employment.

Economic geography is politically important because politicians have varied incentives to cater to more or less geographically concentrated groups depending on a country's electoral institutions. Electoral institutions stipulate the rules governing elections and vary from country

---

[7] In the short- to medium term. In the longer term, they may be able to move depending on their mobility and the costs of adjustment.
[8] Citizens who own factors of production employed in the industry, such as capital or labor, also benefit.

to country. In some democracies, politicians can win office with less than a majority of votes. In others, a candidate's chance of winning office depends not on the number of individual votes they receive but rather on their value in office to party leaders. In short, the path to electoral victory is different in different countries depending on a country's electoral system.

Two main categories of electoral systems exist: plurality and proportional. In a plurality system, votes are cast for individual candidates and the candidate with the most votes wins office (Cox, 1990: 906). In contrast, proportional representation (PR) systems allocate legislative seats to parties in accordance with the proportion of votes won by each party. Together these two formulas govern eighty percent of elections held around the world (Clark, Golder, and Golder 2013, Inter-Parliamentary Union PARLINE database 2013).[9]

I briefly describe how economic geography matters in these two different systems. I develop my argument more fully in Chapter 3 where I identify two mechanisms through which economic geography and electoral institutions shape leaders' incentives and subsequently policy: (1) effective vote maximization and (2) the nature of electoral competition.

Geography in Plurality Systems

Politicians have incentives to cater to geographically concentrated groups in countries with plurality electoral systems because they must win a plurality of votes in their electoral district to win office. Politicians therefore court the support of groups concentrated in their own geographically defined district. To win their support, incumbent politicians provide economic benefits or "rents" to their district. By providing economic benefits to their constituents, politicians seek to develop their own personal support base among voters (i.e. a personal vote).

Politicians can use subsidies to develop a personal vote when the beneficiaries of the subsidy are geographically concentrated in their own district. When an industry's employees are concentrated in a politician's district, subsidies for that industry are analogous to legislative particularism, or "pork." The economic benefits of the subsidy go to the politician's district but the costs are spread over all taxpayers throughout

---

[9] The PARLINE database can be found at www.ipu.org. The remaining 20 percent consist of "mixed" electoral systems that combine features of both plurality and proportional electoral systems. Germany, for example, has a mixed electoral system.

the country. Supplying such geographically concentrated benefits helps politicians cultivate their own personal support base among voters, which increases their reelection chances in plurality electoral systems (Ferejohn 1974, Fenno 1978, Wilson 1986).

Providing subsidies and other economic benefits to geographically concentrated groups is an expedient way to win elections in plurality electoral systems. As a result, particularistic economic policies for geographically concentrated groups are common in countries with plurality electoral systems. In the United States, for example, the Republican-led administration imposed a 30 percent tariff on steel imports in 2002 in an attempt to win Congressional seats in the steel-producing states of Ohio and Pennsylvania (Read 2005).[10] In 2017, President Donald Trump launched an investigation of foreign steel imports in order to fulfil a campaign promise he made to steel workers in two important swing states: Ohio and Pennsylvania. Trump launched the investigation under Section 232 of the Trade Expansion Act of 1962, which empowers the Department of Commerce to decide whether imports "threaten to impair" US national security and gives the president substantial autonomy to impose new trade barriers (Bown 2017). Trade barriers imposed on foreign steel imports would benefit the geographically concentrated steel industry and its employees. At the same time, however, they would increase costs for US manufacturers and construction companies that rely on imported steel inputs, ultimately raising costs for US consumers and also taxpayers who fund public infrastructure projects.

Despite their costs, particularistic economic policies – or even just the promise of them – provide a useful electoral tool in plurality systems when the beneficiaries are geographically concentrated. Particularistic economic policies allow parties to target benefits to precisely those areas where they most need increased voter support, such as Ohio and Pennsylvania in the United States example. In contrast, when the beneficiaries are geographically diffuse, particularistic economic policies are an inefficient means to win plurality elections. If the US steel industry had been more evenly dispersed within the country, for example, providing the industry with economic benefits would have "over bought" support in some states where the Republican party did not need any additional votes to win. For this reason, neither political parties nor individual politicians have strong incentives to provide economic benefits to geographically diffuse groups in plurality systems. For political parties, supporting diffuse groups will over buy support in some areas and under buy support in others. And

---

[10] The US risked violating their obligations as a member of the World Trade Organization by supplying these tariffs to the steel industry.

Argument in Brief

individual politicians seeking an office other than the presidency need to win only the support of voters in their own electoral district. As a result, few incentives exist to work on behalf of geographically diffuse groups spread across many districts in plurality systems because doing so neither sufficiently rewards politicians' efforts nor maximizes their chances for (re)election.

Geography in Proportional Systems

Geography is unimportant for political parties competing in countries with proportional electoral systems and a single, national electoral district. In such systems, all votes are equally valuable because they all contribute to a party's share of the national vote, regardless of their geographic location. A party's national vote share determines how many seats they hold in the legislature. Parties want to maximize the number of legislative seats they hold, and to this end they work to maximize their share of the national vote. They can do so with little regard for the geographic distribution of potential supporters because the entire country constitutes a single electoral district. In reality, however, single district PR system are rare. Only a handful of PR countries have one nationwide electoral district. Instead, most PR systems have multiple subnational districts.

Geography matters in PR systems with multiple electoral districts. In such systems, most legislative seats are awarded to parties based on their district performance rather than their national performance. In Norway, for example, 150 of 169 legislative seats are allocated to parties based on their share of district votes (Aardal 2011).[11] As a result, the geography of potential votes is electorally important in countries with proportional electoral systems and multiple districts. Political parties competing in such countries consequently take economic geography into account when making policy decisions. However, unlike parties in plurality systems, political parties in PR systems tend to favor geographically diffuse groups. Providing economic benefits to geographically diffuse groups maximizes parties' effective votes and the likelihood of being in parliament.

Parties competing in proportional systems with multiple district are better off supplying policies to geographically diffuse groups rather than

---

[11] Because seats are awarded to parties at the district level, we observe disproportionality between parties' national vote shares and the number of legislative seats they hold in most PR countries. Such disproportionality has been the subject of extensive research including, for example, Gallagher (1991).

concentrated groups for several reasons. First, favoring geographically diffuse groups helps parties build a nationwide constituency. A nationwide constituency is electorally useful in proportional systems and particularly in PR systems where elections are party centered. A nationwide constituency helps parties grow their vote share and "displaces the district as the primary electoral constituency" (Lancaster and Patterson, 1990: 470).[12] Displacing the district as the primary electoral constituency gives the party greater influence over their legislators because legislators are less able to appeal to their district-level constituents for reelection. Parties with greater control over their legislators have relatively greater influence on policy outcomes, which permits them greater opportunities to provide benefits to diffuse groups. Such benefits can engender "a shift in the national mood towards the ruling party" (Reed, Scheiner, and Thies 2012), which increases the party's vote share and the number of seats they control in the legislature.

Second, political parties have incentives to pursue the support of geographically diffuse groups to ensure that the party's vote share is above any national vote-share threshold, which exist in many PR systems. These thresholds stipulate that political parties must win a minimum share of the national vote to hold any seats in parliament. Parties that pursue the support of geographically concentrated groups rather than geographically diffuse group may fail to cross national vote-share thresholds. In Norway, for example, a party called People's Action Future for Finnmark (*Folkeaksjonen Framtid for Finnmark*) focused exclusively on improving the economic conditions in Finnmark. To this end, the party campaigned on increasing government assistance for the area's fishing industry (Aardal 2011). The party subsequently won 21.5 percent of the vote in the electoral district of Finnmark in 1989 (Aardal 2011). However, the party won just 0.3 percent of the national vote and as a result it was not eligible for any of the legislative seats allocated at the national level because it failed to clear the national threshold of 4 percent. As this example illustrates, parties competing in PR systems have compelling incentives to pursue diffuse votes spread across the country.

Third, parties in PR systems may support geographically diffuse groups in an attempt to generate a more uniform vote swing – that is, a similarly sized vote increase in all districts. A more uniform swing often produces more seats for parties competing in PR systems with multiple districts. Because a more uniform swing potentially increases a party's legislative

---

[12] "Thereby decreasing the importance of pork-barrel politics" (Lancaster and Patterson, 1990: 470).

seats, parties have electoral incentives to favor geographically diffuse groups.

These reasons explain why geographically diffuse groups receive support from political parties competing in PR systems. But why would parties in PR systems provide fewer subsidies to concentrated groups than diffuse groups? The answer is simple: subsidies entail costs. These costs include both real monetary costs and opportunity costs. Every dollar spent on subsides for concentrated groups is one less dollar available for diffuse groups. The opportunity costs of forgone spending on diffuse groups are large for political parties competing in PR systems. Subsidizing the geographically diffuse construction industry, for example, helps people in all regions of the country. Employees in the construction industry benefit directly from subsidies via increased wages and more secure employment. Owners of capital invested in the industry benefit from above market rates of return and greater demand. Related sectors, such as real estate and retail, also benefit from government-funded subsidies for the construction sector. And because the sector is geographically diffuse, many more people in related sectors indirectly benefit from government support. Real estate agencies, restaurants, and hardware stores across the country benefit from government aid to the diffuse construction sector. If the sector was concentrated in a single area, many fewer people in related sectors would benefit – potentially just a handful of real estate agents, restaurants or hardware stores in a single city. More people, in more places, benefit from subsidies to geographically diffuse groups.

Subsidies to geographically diffuse industries typically benefit more people than subsidies to equally sized concentrated industries. In effect, there is a "dispersion bonus" from subsidizing geographically diffuse industries.[13] This dispersion bonus is more valuable electorally for parties competing in PR systems than parties in plurality systems. In PR systems, every additional vote won by a party contributes to its electoral success. In contrast, many of the additional votes secured via subsidies are lost to parties and politicians competing in plurality systems. As a result, governments in PR systems will tend to spend more on geographically diffuse groups than governments in plurality systems, all else equal. Even relatively small, geographically diffuse groups can win subsidies in PR systems because more votes translate into more seats. In Norway, for example, the Liberal Party (*Venstre*) could have won seven seats instead of two if it had won just 0.1 percent more of the national vote in the 2009 election (Aardal 2011). Providing subsidies to an industry employing just

---

[13] I am grateful to John Carey for articulating the term "dispersion bonus."

0.1 percent of the country's labor force could have made a big difference to the Liberal Party's electoral fortunes.

Because the electoral support of geographically diffuse groups is especially valuable for parties competing in PR systems, diffuse groups can and do win particularistic economic policies. In Sweden, a country with a proportional electoral system, the geographically diffuse forestry sector received 10 percent of all government subsidies – despite the fact that it employed less than 1 percent of the country's total population (Carlsson 1983).[14] Similarly, in Norway, which also has a PR system, the geographically diffuse tourism industry receives generous state support. In 2013, for example, the Norwegian government made a deal with Disney regarding the marketing of the film *Frozen*. For an undisclosed amount of money, the Norwegian government secured exclusive rights to use creative elements from the film, as well as the Disney logo and brand, in the marketing of Norway as a travel destination (Innovation Norway 2014a). The deal is credited with significantly increasing tourist numbers. Fjord Tours' sales in the American market doubled in the beginning of 2014, and ticket sales on the Hurtigruten coastal express increased by 24 percent (Innovation Norway 2014a). The upsurge in tourism brought economic benefits to businesses throughout the country.

In sum, my argument brings together electoral institutions and economic geography and shows how they interact to shape economic policy. I argue that both plurality and proportional electoral systems incentivize the provision of narrowly beneficial, particularistic economic policies under certain conditions. Leaders in plurality systems have incentives to supply particularistic economic policies when the beneficiaries are geographically concentrated. When the beneficiaries of particularistic economic policies are geographically diffuse, leaders in PR systems have incentives to supply such policies.

CONTRIBUTION

By bringing together geography and institutions, my argument provides a solution to the ongoing debate over which democratic institutions make governments most responsive to narrow interests. Purely institutional accounts reach conflicting conclusions about the effects of electoral systems on leaders' responsiveness to narrow interest groups. In the following section, I briefly outline the contours of the ongoing debate.

---

[14] During the mid-1970s.

Conventional Wisdom About Plurality Systems

Many scholars believe plurality electoral systems provide the greatest incentives for politicians to cater to narrow interests. Grossman and Helpman (2005) develop a theoretical model that illustrates why plurality systems generate incentives for policy targeting. In their model, two parties compete in legislative elections, and each party has equal chances of winning a given seat in a given district. There are three electoral districts; each district contains one-third of the population and elects one legislator. Grossman and Helpman (2005) assume that for each industry, all capital is owned by the residents of a single district. As a result, legislators represent constituents with industry-specific economic interests.

Upon forming the government, the delegation from the majority party seeks to maximize the welfare of its constituents. If the party in power represents all three districts, the legislature will work to maximize the welfare of the entire country by setting tariffs at zero. In contrast, if the governing party represents only two of the three districts, they will set a positive tariff rate. Tariffs typically provide narrow benefits to select domestic producers – often at the expense of consumers.[15] A nonzero tariff emerges in Grossman and Helpman's model when trade policy is chosen by the majority delegation. Legislators in the majority use tariffs to redistribute income to residents in their own districts, rather than maximize national welfare by setting an optimal tariff of zero.

The model generates expectations about the effects of electoral systems because Grossman and Helpman (2005) speculate that the governing party in majoritarian systems is unlikely to represent all three districts. Instead, the government in a proportional electoral system is far more likely to represent all three districts. Thus the legislature in a proportional system will work to maximize the welfare of the entire country by setting tariffs at zero. In other words, the Grossman and Helpman (2005) model suggests that PR systems will produce more broadly beneficial economic policies. Majoritarian systems, in contrast, will lead politicians to supply more particularistic economic policies, such as industry-specific tariffs or subsidies.

A similar prediction emerges from a canonical model developed by Persson, Roland, and Tabellini (2007). They assume that the geographic

---

[15] Of course, tariffs can be designed to be more or less narrowly beneficial. In theory, a tariff could be designed to protect a product produced by just one firm. Such a tariff would provide narrow benefits. At the other extreme, a government could impose a flat tariff that taxes all imported goods at the same rate, as was historically the case in Chile.

distribution of economic groups is the same in all districts (Persson et al. 2007). Given this assumption, their model predicts that majoritarian elections will be associated with more narrowly beneficial policies and less broadly beneficial public goods, as compared to proportional elections. Persson and Tabellini find empirical support for this prediction; they report that majoritarian systems, a type of plurality electoral system, produce smaller government and less welfare spending than proportional elections (Persson and Tabellini, 1999: 2003). Spending on social services is estimated to be 2 to 3 percent lower in plurality systems, as compared to proportional systems, holding all else equal (Persson and Tabellini 2003).

Several additional studies report similar findings. In a sample of 147 countries from 1981 to 2004, Evans (2009) finds that majoritarian systems have higher tariffs than countries with proportional electoral systems, all else equal. Ardelean and Evans (2013) demonstrate that tariffs are higher, on average, in majoritarian systems than in proportional systems using product-level tariff rates for a cross-section of developed and developing countries between 1988 and 2007. Also, democracies with plurality electoral rules are named as defendants in international disputes over illegal narrow trade barriers more often than democracies with proportional electoral rules (Rickard 2010, Davis 2012). This evidence suggests that the electoral incentives to provide narrowly beneficial policies are so compelling in plurality systems that legislators are willing to supply such policies even when doing so violates their international treaty obligations (Naoi 2009, Rickard 2010, Davis 2012). This evidence supports the proposition that plurality electoral systems make governments highly responsive to special interests.

Conventional Wisdom About Proportional Systems

Some scholars argue that the incentives to cater to narrow interests are even *greater* in proportional systems compared to plurality systems. A formal model developed by Rogowski and Kayser (2002) highlights the importance of seat-vote elasticities. Majoritarian systems have greater seat-vote elasticities than PR systems, and as a result, a loss of votes translates into a greater loss of seats for parties competing in majoritarian systems. In proportional systems, politicians are able to cater to narrow interests without having to be overly concerned with any election losses they might incur for doing so. In contrast, politicians in plurality systems cannot stray far from the preferences of the median voter because a small change in vote share can produce a large change in seat share. Rogowski and Kayser therefore posit that politicians in proportional systems will be relatively more responsive to narrow interests, such as industry-specific demands for protection.

A theoretical model developed by de Mesquita et al. (2003) also implies that narrow, particularistic policies will be more frequent in PR systems, as compared to majoritarian systems. Their model examines the political consequences of the size of a "winning coalition." A winning coalition is a subset of the selectorate (i.e. the people who choose the leaders) with sufficient size to allow the subset to endow leaders with political power to negate the influence of the remainder of the selectorate and the disenfranchised members of the society (de Mesquita et al. 2003: 51). The winning coalition is larger in majoritarian systems than in PR systems, according to de Mesquita et al. (2003).[16] As the size of the winning coalition grows, the cost of private goods, such as subsidies, increase. According to their logic, narrow policies like industrial subsidies should be less frequent in majoritarian systems than in proportional systems.

Empirical evidence exists to support the idea that proportional electoral rules incentivize particularistic economic policies. Nontariff barriers are higher, on average, in proportional rule democracies than in majoritarian systems (Mansfield and Busch 1995). Proportional rule systems are also associated with higher consumer prices (Rogowski and Kayser 2002, Chang, Kayser, and Rogowski 2008, Chang et al. 2010). Higher consumer prices may reflect governmental policies that privilege producer groups at the expense of consumers. Legislatively imposed barriers to trade, for example, raise the prices of consumer goods. The existence of higher trade barriers in proportional rule countries may explain why consumer prices are higher in PR systems than in majoritarian systems. This evidence stands in direct contrast to the results reported in several other studies as described above and illustrates the lack of consensus over the effects of electoral systems on economic policy. While some scholars believe that PR systems make leaders especially responsive to special interests and consequently generate more particularistic economic policies, others believe that plurality systems do. The debate has continued unresolved to date.

Going Forward: Geography

My argument suggests a solution to the current impasse: economic geography. Up to now, geography has been largely absent from discussions of electoral institutions' policy effects. Some studies overlook geography entirely (e.g. Persson et al. 2007). Others make

[16] However, Persson and Tabellini (2003) make the opposite claim.

strong assumptions about the geographic distribution of economic activity. Some models assume that each electoral district contains one unique industry that is entirely concentrated within the district (e.g. McGillvray, 1997: 588, 590, Grossman and Helpman 2005, McGillivray 2004). Such restrictive assumptions bear little resemblance to reality. Although economic activity is "lumpy," industries are rarely contained within a single electoral district. Moreover, different industries and economic sectors display different patterns of geographic dispersion, and these patterns vary between countries.

Ignoring the geographic patterns of voters with shared economic interests may be innocuous if politicians elected via different rules are equally responsive to concentrated (or diffuse) interests. But different electoral systems generate different incentives to represent geographically concentrated (or diffuse) groups. Therefore, geography must be part of any institutional explanation of policy outcomes. Identical institutions can produce different policy outcomes depending on countries' economic geography. Plurality systems will produce generous subsidies only when the potential beneficiaries are geographically concentrated. When the beneficiaries are geographically diffuse, subsidies will be relatively meagre in plurality systems. In this way, alike electoral institutions can produce different policy outcomes depending on the pattern of economic geography. Taking geography into account can help to explain the mixed effects of electoral institutions on policy outcomes.

While I am not the first to suggest the importance of geography for politics,[17] virtually all existing research focuses exclusively on plurality systems. McGillivray (2004) examined the effects of geographic concentration in two plurality countries: Canada and the United States. McGillivray found that concentrated industries in these two countries tend to win more trade protection than diffuse industries. Similarly, Hansen (1990) established that geographically concentrated industries are more likely to secure protection from foreign import surges in the United States. Milner (1997) showed that concentrated industries in the United States made fewer trade concessions in negotiations over the North American Free Trade Agreement (NAFTA).

Although geographic concentration is politically advantageous for interest groups in plurality systems, it is unclear what role geography plays in proportional electoral systems.[18] Cognizant of this limitation,

---

[17] See, for example, Rodden (2010).
[18] Busch and Reinhardt (2005) argue that geographic concentration may be an asset for all interest groups regardless of a country's electoral rules.

McGillivray (1997) recommended future research to investigate the effects of geographic concentration in proportional systems (p. 604).[19] My research responds to her appeal by examining the effects of geographic concentration in countries with various different electoral systems. This book provides one of the first quantitative tests of the policy effects of economic geography in countries with proportional electoral systems. I find that the effects of geographic concentration vary depending on a country's electoral institutions. While geographic concentration is a political asset for interest groups in plurality systems, it is a political liability in proportional systems. In other words, the political consequences of economic geography vary between electoral systems.

A key lesson of this book is that economic geography matters for politics and policy. It consequently merits greater attention from academics and policy makers alike. Economic geography has important political consequences – many of which have been overlooked to date. Economic geography is credited by some with the election of Donald Trump to the US presidency (Autor et al. 2016) and Britain's decision to leave the European Union in 2016 (Colantone and Stanig 2016). Voters' decisions appear to be influenced, at least in part, by their regions' experience with globalization. Some regions lost out to globalization while others grew rich because of the unequal distribution of economic activities across space within countries. Regions differed in their exposure to foreign import competition because the importance of different industries for local employment varied between regions (Autor et al. 2013). Regions hit hard by foreign imports experienced decreases in wages and employment (Autor et al. 2013). People living in economically depressed regions disproportionally supported Trump in the 2016 US presidential election and voted to leave the EU in the UK's 2016 referendum on exiting the European Union. These examples suggest that economic geography matters for voters' decisions at the ballot box, and as a result can have far reaching consequences.

Economic geography has also been credited with the recent rise in populism. The growing appeal of populist politics is believed to be due, in part, to increasing regional inequalities. Among OECD countries, the average productivity gap between the most productive 10 percent of

---

[19] Although McGillivray (2004) hypothesizes about the effects of district marginality in both PR and plurality countries, her empirical tests include only plurality countries. Her sample does not extend to PR countries (McGillivray, 1997: 271, McGillivray, 2004: 81). McGillivray herself writes, "The hypotheses for proportional representation systems are not examined" (2004: 87).

regions and the bottom 75 percent widened by nearly 60 percent over the past 20 years.[20] Rich regions are increasingly pulling away from less well-off regions. In Britain, for example, London's share of the country's gross value added rose from 19 percent to 23 percent during the period from 1997 to 2015.[21] Growing regional inequalities may help to explain the rise in populism and highlight the political significance of economic geography. This book makes an important contribution to understanding the political consequences of economic geography.

WIDER IMPLICATIONS

I develop my argument in the context of economic policy. However, the logic of my argument is general and can be applied to other issue areas. Whenever voters with shared preferences exhibit varied geographic patterns, my argument can provide useful insights. One example may be ethnic politics (Horowitz 1985, Mozaffar, Scarritt, and Glalich 2003, Lijphart 2004). My argument suggests that the influence of an ethnic group's shared policy preferences will depend on the country's electoral institutions and the group's geographic distribution. When an ethnic group is geographically diffuse, their preferences will have greater expression under proportional electoral rules than plurality rules. In this situation, it may be inopportune to introduce plurality electoral rules – particularly in an ethnically diverse society.[22]

Understanding that institutions alone do not determine policy has further implications for constitutional designers and reformers. When reformers deliberate over how to alter their country's constitution, they frequently focus on how to achieve desired policy outcomes. However, institutions alone cannot guarantee any specific outcome. Instead, policy outcomes depend on both institutions and economic geography. When designing institutions with particular policy goals in mind, leaders must consider the geographic distribution of citizens with shared interests in key policy goals.

My argument also has several implications for countries' international economic relations. First, my argument suggests when and under what conditions countries are most likely to violate international agreements that restrict the use of particularistic economic policies. Narrowly targeted policies are the focus of many international agreements because they tend to cause economic distortions (Rickard 2010). International

---

[20] *The Economist* October 21, 2017: 20.
[21] *The Economist* October 21, 2017: 20.
[22] Of course, future research is needed to determine the extent to which my argument applies to ethnic politics.

agreements seek to limit distortions by regulating governments' use of particularistic economic policies, such as industry-specific tariffs and subsidies. My argument suggests that democracies with plurality electoral systems and geographically concentrated industries have powerful domestic incentives to violate these international restrictions. Similarly, democracies with proportional electoral systems and geographically diffuse industries have powerful domestic incentives to violate international restrictions on particularistic economic policies.

Second, my argument suggests which countries are most likely to impede future international economic cooperation and why. Governments elected via plurality rules and facing demands for subsidies from geographically concentrated groups, have strong incentives to resist new international agreements that restrict subsidies. The Canadian government, for instance, nearly scuppered the Trans-Pacific Partnership (TPP) agreement – an agreement negotiated by twelve countries over five years that would cover nearly 40 percent of global trade – to protect subsidies for thirteen thousand Canadian dairy farmers.[23] This small group of farmers enjoyed such influence because of their geographic concentration and Canada's plurality electoral system. This example serves as a reminder that all politics is local. Countries' international economic relations are shaped by the interactive effect of domestic political institutions and economic geography. In short, domestic politics influence international economic relations.

WHY SUBSIDIES?

Although the logic of my argument is general, I test the hypotheses derived from my theory using data on government-funded subsidies. Government spending on subsidies reveals how leaders weigh the demands of special interests against general welfare. Subsidies typically benefit smaller groups at the expense of larger groups, such as taxpayers. Subsidies to the US sugar industry, for example, sustain a domestic sugar price two to three times higher than the world's market price. As a result, approximately 20,000 US sugar cane farmers receive an extra $369 million dollars a year (Beghin et al. 2003, Frieden, Lake, and Schultz, 2010: 234).[24] These benefits come at a cost to American

---

[23] The finalized TPP proposal was signed by twelve countries. However, the twelve countries must complete their respective domestic treaty-ratification processes before the agreement can come into force. In 2017, President Donald Trump withdrew the US from the Trans-Pacific Partnership.

[24] Above the internationally determined value for the commodity. Calculations are for 1998 converted into 2006 US dollars.

taxpayers and consumers who pay an additional $2.3 billion dollars a year for sugar (Beghin et al. 2003, Frieden et al., 2010: 234). In this way, subsidies redistribute wealth from taxpayers to producers – that is, from larger groups to smaller groups. Governments' decisions about subsidy spending reveal their willingness to support narrow interest groups at the expense of broader groups.

Surprisingly little is known about the politics of government-funded subsidies.[25] On their face, subsidies are a type of distributive policy. They involve taxes and transfers and necessitate decisions about the allocation of government assistance to identifiable localities or groups (Golden and Min 2013). Given this, subsidies may share many of the characteristics of distributive policies. Yet, different distributive policies benefit various different groups and as a result engender dissimilar politics.

The beneficiaries of subsidies vary. Some subsidies are selectively targeted by governments to a few producers using strict eligibility criteria. For example, one of the subsidy programs examined in Chapter 5 assists only Cognac producers located in the French regions of Charente and Charente-Maritime. In contrast, other programs assist virtually all producers in a country with few, if any, qualifying criteria. In Austria, for example, the government subsidized the purchase of microwaves by restaurants across the country. Given the diversity in subsidy recipients, subsidy programs cannot be classified as being "narrowly beneficial" or "broadly beneficial" without knowing more about the recipients themselves. Knowing who benefits from a subsidy and where the beneficiaries are located geographically helps to elucidate the policies behind subsidy programs.

Understanding the politics of subsides is important because government-funded subsidies have potentially serious implications for redistribution and inequality. Governments spend large amounts of money on subsidies. In the European Union, subsidies to the manufacturing sector accounted for 2 percent of value added or approximately €1,000 per person employed in the sector (Sharp 2003). In the United States, Fortune 500 corporations received more than 16,000 subsidies, worth $63 billion dollars (Mattera and Tarczynska 2015). And many governments continue to spend generously on subsidies even during periods of fiscal austerity (Sherman 2017).[26] The Australian government increased spending on manufacturing subsidies by 26 percent from 2006-2010.[27] Money spent

---

[25] Although see Blais (1986), Verdier (1995), Alt et al. (1999), Zahariadis (2001), and Aydin (2007).
[26] http://www.baltimoresun.com/business/bs-bz-business-incentive-spending-grows-20170302-story.html
[27] Author's calculations from Australian budget data available at www.budget.gov.au/past_budgets.htm.

on subsidies is money that is no longer available for other programs, such as welfare or health care (Rickard 2012b). As one government employee succinctly put it, "every subsidy I am giving is the money that the government could have spent elsewhere. Every subsidy means a primary healthcare centre I cannot build."[28] In short, subsidies have important implications for the regressivity of government spending.[29]

More and more governments use subsidies to aid domestic producers. Traditionally, many governments used tariffs to assist domestic businesses. But tariffs are restricted by a growing number of international trade agreements, and as a result, many governments have turned to subsidies instead (Ford and Suyker 1990, OECD 1998, Rickard 2012b). The Japanese government, for example, planned to increase subsidies to pig farmers to compensate them for the reduction in pork tariffs established as part of the Trans-Pacific Partnership (TPP) multilateral trade agreement.[30] Because international agreements increasingly restrict the use of tariffs and efficient capital markets restrict exchange rate manipulation, subsidies are one of the few tools policy-makers still have at their disposal to help domestic producers (Thomas 2007).[31] Subsidies are therefore becoming an increasingly common mode of state intervention in industry (Verdier 1995).

In many developed economies today, governments face intense pressure to assist the declining manufacturing sector. Support for manufacturing was a key issue in the 2016 US Presidential Election campaign, for example. Trump's promise to bring "manufacturing back to America"[32] helped him win votes in traditionally Democratic areas, which were crucial to his victory. In the United Kingdom, Prime Minister Theresa May created a new government ministry charged with developing

---

[28] Bibek Debroy, economist at the National Institution for Transforming India, a Government of India policy think-tank. Quoted in the Economic Times, India Times, May 7, 2017, "Scrap all production subsidies" http://economictimes.indiatimes.com/news/economy/policy/scrap-all-production-subsidies-niti-aayogs-bibek-debroy/articleshow/58560312.cms.

[29] Subsidies themselves may be regressive; they are funded by taxes but often go to support rich, powerful groups, such as US sugar cane growers.

[30] The Japan News, May 21, 2014, S Edition, Business Section, p. 8. Accessed via Lexis Nexis. Currently a tariff of up to ¥482 yen is imposed on one kilogram of cut meat, such as pork tenderloin or pork loin. However, the figure was to be gradually reduced to ¥50 yen over fifteen years as part of the TPP agreement.

[31] Some international restrictions on subsidies exist. However, these restrictions tend to be more lax than international restrictions on tariffs (Rodrik 2004). Furthermore, it is often more difficult to determine if governments are subsidizing producers in violation of international rules because of subsidies' relative opacity (Kono 2006).

[32] www.whitehouse.gov/the-press-office/2017/02/23/remarks-president-trump-meeting-manufacturing-ceos.

an industrial policy to assist the manufacturing sector. The new ministry's title included the words "industrial strategy," sending a clear message that the United Kingdom is not shy about having a proactive industrial policy (Pratley 2016).[33] The minister appointed to head the department said he had been "charged with delivering a comprehensive industrial strategy" by the prime minister (Ruddick 2016).

Industrial policies – and subsidies for the manufacturing sector in particular – appear to be gaining political importance (Stöllinger and Holzner 2016). Questions about how best to adjust to the competitive pressures of a global economy have put subsidies at the center of contemporary policy discussions. In 2017, for example, the World Bank published a new working paper on subsidies and industrial policy (Maloney and Nayyar 2017). As governments come under growing pressure to assist firms, industries, and sectors, understanding the politics of subsidies is more important than ever. The theory developed in this book explains which countries' governments respond most vigorously to businesses' demands for economic assistance and why.

Despite the growing importance of subsides, governments in different countries spend different amounts of money on subsidy programs. Governments in the United Kingdom, for example, often provide subsidies to individual firms (Sharp et al. 1987). In fact, the Industry Act of 1972 explicitly authorized governments to subsidize individual firms and industries in order to boost investment (Bailey 2013). Throughout the 1970s, the British government subsidized individual firms, including the state-owned companies British Steel and British Airways (Sharp et al. 1987). In contrast, during the same period, the West German government rarely supplied particularistic economic policies (Owen 2012). Instead, the government's policy portfolio consisted principally of programs that benefited large groups of citizens (Owen 2012). Subsidies, when provided, were made available to all industries rather than a select few (Shepherd, Duchêne, and Saunders 1983). The government routinely refused to provide subsidies to individual firms (Schatz and Wolter 1987). It focused instead on building a comprehensive framework of policies that would benefit large numbers of citizens called the *Soziale Marktwirtschaft* (Sharp et al. 1987).

Cross-national variation in subsidies persists today, despite ever increasing global economic integration. For example, France typically spends three to four times as much on subsidies as Italy. In 2013, French subsidies equaled 0.6 percent of GDP while Italian subsidies totaled just

---

[33] The Ministry's full title is the Department for Business, Energy and Industrial Strategy.

0.2 percent of GDP. Such variation is surprising because globalization is believed to compel governments to adopt similar economic policies. Governments in a given country may come under pressure to subsidize domestic business in the face of foreign subsidies. Subsidies to China's steel industry, for instance, prompted demands for government assistance in other steel-producing countries including France, the United States, Germany, and the United Kingdom (Wong 2014, Rankin 2016, Farrell 2016). Despite competitive pressures from increased international trade and foreign subsidies, countries continue to exhibit varied levels of support for domestic producers.

Even in the wake of the 2008 global economic crisis, governments' enthusiasm for particularistic economic policies varied between countries. Some leaders raced to provide aid selectively to troubled industries. The French president unveiled a "strategic national investment fund" to buy stakes in certain French industries to protect them against foreign "predators."[34] In Italy, the Prime Minister called government-funded subsidies a "categorical imperative" in times of economic distress.[35] The Australian government increased budgetary assistance to industry by 26 percent from 2006 to 2010 and the British government poured money into select parts of the economy that were deemed to "make a difference."[36] But not all governments were equally keen to help select domestic producers. Governments in Germany and Sweden refused to selectively assist ailing industries in the aftermath of the 2008 crisis. The cross-national variation in government-funded subsidies highlights the key puzzle motivating my research: why are democratically elected leaders in some countries more willing to supply particularistic economic policies than others?

EMPIRICAL STRATEGY

To better understand the interactive effect of electoral institutions and economic geography on economic policy, I examine three different types of variation in subsidies. First and foremost, I examine the variation in government spending on subsidies between countries, holding all else equal. Second, I examine the variation in government spending on subsidies within countries between sectors – exploiting the disparity in employment patterns in different sectors of countries' economies. Third, I examine the variation in subsidy spending between electoral

---

[34] *The Economist* November 1, 2008: 62.
[35] *The Economist* November 1, 2008: 62.
[36] *The Economist* November 1, 2008: 62 and author's calculations from Australian budget data available at www.budget.gov.au/past_budgets.htm.

districts within countries, holding economic geography constant. All three tests highlight the importance of the incentives politicians and political parties face as a result of the interaction of electoral institutions and economic geography.

PLAN OF THE BOOK

In Chapter 2, I describe economic geography in greater detail and explore how and why it varies within and between countries. In Chapter 3, I develop my argument in full and identify two mechanisms through which economic geography and electoral institutions work together to shape leaders' incentives and subsequently policy: (1) effective vote maximization and (2) the nature of electoral competition.

In Chapter 4, I test my argument empirically using large-N, quantitative methods. I construct a politically relevant measure of economic geography using disaggregated employment data. These data identify employees' geographic location as well as their sector of employment and are available for more than a dozen economically advanced countries over two decades. This empirical measure obviates the need for simplifying assumptions about economic geography. Instead, actual patterns of economic geography can be measured in a cross-nationally comparable fashion.[37] These data demonstrate that employment patterns rarely conform to the restrictive assumptions in many prominent models of economic policy making. Relaxing these assumptions reveals new predictions about electoral institutions' policy effects and provides a novel solution to the ongoing debate over which institutions generate the most particularistic economic policies.

Statistical tests of governments' subsidy spending in economically developed democracies over nearly two decades reveal that economic geography conditions the effects of electoral institutions – controlling for international subsidy rules, trade openness, country size, economic development, and various other features of a country. Subsidies for manufacturing constitute a larger share of government expenditures in plurality systems than in PR systems when manufacturing employment is geographically concentrated. When manufacturing employment is diffuse, governments in PR systems assign relatively more of their budgets to manufacturing subsidies than governments in plurality systems, holding all else equal. These results are robust to alternative model specifications including those that relax the assumption that electoral systems are exogenous.

---

[37] Albeit for a limited number of countries given the highly disaggregated data needed.

In Chapter 5, I examine the mechanisms linking electoral institutions and economic geography to economic policy outcomes via two case studies. Given the ubiquity of government subsidies, it would be easy to cherry pick cases that fit my theory. To guard against this, I use a methodical, multistep selection criterion. Two subsidies meet the comprehensive criterion: a program to support Cognac producers in France and a subsidy for farm-gate wine merchants (i.e. wine makers who sell their wine at the place of production) in Austria. Both subsidies conform to my theoretical expectations. In France, where legislators compete in single-member districts via a majority-plurality electoral system, subsidies tend to be targeted to geographically concentrated producers – as in the Cognac example. In contrast, Austrian subsidies tend to go to geographically diffuse groups. In Austria, legislators compete in party-centered elections in multimember districts where legislative seats are awarded via proportional rules and party leaders determine the identity of the candidates that fill the party's seats. Given these institutions, I expect Austrian politicians will be relatively more responsive to the interests of geographically diffuse groups. Consistent with this expectation, geographically diffuse farm-gate wine merchants won generous economic benefits from the Austrian government.

Both Austria's and France's subsidies violated the European Union rules on State Aid, which limit member-states' ability to subsidize domestic producers. The European Union (EU) and World Trade Organization (WTO) regulate subsidies in an attempt to minimize economic distortions and create a level playing field for businesses in all member countries. The international restrictions on subsidies may make it more difficult to observe the domestic politics behind such programs. For instance, governments with electoral incentives to provide subsidies may not do so because of international rules limiting subsidies. Such behavior would make it difficult to find support for my argument; I would not see subsidies where I expect to, given a country's electoral institutions and economic geography. In this way, international restrictions on subsidies bias against finding support for my argument.

In practice, however, many governments continue to subsidize domestic producers – even as members of the EU and WTO (OFT 2009). Most international restraints on subsides are "either voluntary or do not bind in a significant way" (Rodrik, 2004: 32). Governments still have "much scope" to subsidize domestic producers (Rodrik, 2004: 32). Most international rules provide general block exemptions that allow for certain types of subsidies. Subsidies are permitted for research and development, innovation, training, employment, environmental protection, as risk capital, and for promoting entrepreneurship (OFT 2007). The EU state aid rules also allow governments to subsidize failing firms (European Commission 2009).

Government make use of these exemptions to subsidize domestic producers. Governments do so when a country's electoral institutions and economic geography align to provide compelling electoral incentives to fund subsidies. In Chapter 5, I show that a country's electoral institutions and economic geography robustly predict the probability that a government will violate EU state aid rules. In other words, the same variables that predict the generosity of government-funded subsidies also predict states' (non)compliance with international subsidy rules.

In Chapter 6, I investigate subsidies in countries with proportional electoral systems. Among proportional systems, two key institutional features vary: list type and district magnitude. District magnitude refers to the number of candidates elected to parliament from each district. List type determines the order in which a party's candidates receive seats in the legislature. List type influences the nature of electoral competition. In closed-list systems – where voters select a party at the ballot box – elections are party-centered. In contrast, open-list systems allow voters to select individual candidates from a party's list and elections are consequently candidate-centered. The nature of electoral competition shapes the optimal (re-)election strategy of candidates and parties and consequently their policy priorities.[38] As a result, spending on subsidies for geographically diffuse groups is relatively higher in closed-list PR systems, as compared to open-list PR systems. In open-list systems, some funds are diverted away from diffuse groups by powerful individual legislators to assist voters in their own electoral district or bailiwick.[39] Legislators in open-list PR systems have incentives to divert resources in this way to cultivate their own personal support base. Doing so helps them win more individual votes and consequently increases their reelection chances. As a result, spending on subsidies for geographically diffuse groups is relatively higher in closed-list PR systems where political parties can better discipline their legislators to ensure that subsidies flow to geographically diffuse groups.

---

[38] The nature of electoral competition may also interact with district magnitude to shape leaders' incentives (e.g. Carey and Shugart 1995, Shugart, Valdini, and Suominen 2005, Chang and Golden 2007, Carey and Hix 2011).

[39] For evidence of this dynamic see Ames (1995) and Golden and Picci (2008).

In the second half of Chapter 6, I investigate the variation in subsidies to different sectors in a PR country. This single-country study holds constant institutional features, such as electoral rules and list type, as well as other country-specific factors, like culture. Doing so helps to isolate the effects of economic geography. The geographic patterns of employment vary between sectors in a country's economy. I find that geographically diffuse sectors win more generous subsidies than concentrated sectors in a PR country with de facto closed party lists (Norway), holding all else equal. This finding provides further evidence that political parties reap electoral benefits from subsidizing geographically diffuse groups in proportional electoral systems.

In Chapter 7, I examine the variation in subsidies between electoral districts within a country, controlling for economic geography. Specifically, I investigate the cross-district variation in subsidies in Norway – an archetypal PR country. This is a particularly useful case study because most existing research on particularistic economic policies focuses exclusively on plurality countries. As a result, far less is known about the politics of policy targeting in countries with proportional electoral rules. Like most PR countries, Norway has multiple electoral districts, and most legislative seats are awarded to parties based on their share of district-level votes rather than the national vote. As a result, district-level competitiveness may shape parties' election strategies and policy priorities.

Two novel results emerge from an investigation of the variation in subsidies between electoral districts in a PR country. First, political parties competing in PR systems with multiple electoral districts engage in policy targeting (i.e. they supply benefits to select, geographically defined areas). Second, political parties target economic benefits disproportionality to districts where they have relatively more supporters in this PR system. Both findings run counter to the conventional wisdom regarding distributive politics, which is derived largely from studies of plurality systems.

In Chapter 7, I report qualitative evidence from interviews of government ministers and bureaucrats responsible for the administration of subsidy programs in Norway. The interview evidence confirms the importance of electoral politics for governments' spending decisions. Usefully, the interviewees describe how policy targeting works in practice – highlighting the various mechanisms that government ministers and political parties use to control bureaucrats and target subsidies to politically important areas.

In sum, I examine three different types of variation in government-funded subsidies: variation between countries, variation between sectors within countries, and variation between electoral districts within countries. All three investigations illustrate the importance of electoral incentives in shaping governments' economic policies. The incentives to provide selective economic benefits depend on both a country's electoral institutions and economic geography. Building on these findings, the final chapter extends the implications of the argument for a broader theory of policy-making in an increasingly globalized world.

2

The Uneven Geographic Dispersion of Economic Activity

One of the most striking features of "real-world economies" is the uneven distribution of economic activity (Krugman, 1990: 1). Activities, such as production and employment, are "lumpy" – that is unevenly distributed across space – within countries. As a result, some regions remain stubbornly undeveloped while others grow steadily, attracting ever more firms, capital, and people (Brülhart and Traeger, 2005: 598). In Britain, for example, London's economy has grown markedly over the past several decades while other regions, particularly those in the north of England, have declined. As a result, London's share of the economy's gross value added rose from 19 percent to 23 percent over the period from 1997 to 2015.[1] Similarly, in the United States, the bulk of the population resides in a few metropolitan areas, although much of the country's land is fertile, cheap, and sparsely populated (Krugman, 1990: 1). Even in smaller countries, economic activity is often distributed unevenly across space.

The uneven distribution of economic activity has important economic and political consequences. Because economic activity is lumpy, the costs of globalization fall disproportionately on some communities. Communities differ in their exposure to import competition as a result of regional variation in the importance of manufacturing for local employment (Autor et al. 2013). In regions where manufacturing is relatively more important for local employment, rising foreign imports result in higher unemployment, lower labor force participation, and reduced wages (Autor et al. 2013). But while some regions decline, at the same time others prosper. Regions less dependent on manufacturing often benefit from increased international economic integration. Regions' varied economic fortunes influence politics. In the United States, for example, voters in congressional districts exposed to larger increases in foreign imports disproportionately removed moderate politicians from

---

[1] *The Economist* October 21, 2017: 20.

office in the 2000s (Autor et al. 2016). In the United Kingdom, regions exposed to greater inflows of goods from China voted to leave the European Union at higher rates in the 2016 referendum on EU membership (Colantone and Stanig 2016).

Despite the political consequences of economic geography, many scholars fail to take geography seriously. Some scholars simply ignore the uneven distribution of economic activity. Persson, Roland, and Tabellini (2007) explicitly assume that "the distribution of economic groups is the same in all districts" (p. 11). Others assume that economic geography correlates perfectly with political boundaries. In Grossman and Helpman's (2005) model, each electoral district contains one unique industry by assumption. Similarly, McGillivray (1997) assumes that concentrations of industries occur entirely within separate, geographically defined electoral districts (p. 588, 590). And Cassing, McKeown, and Ochs (1986) explicitly assume that each region contains only one industry.

These assumptions bear little resemblance to observed patterns of economic geography. In reality, economic activities vary in their geographic dispersion, and this variation rarely corresponds with politically relevant boundaries. Although some economic activities are strikingly concentrated, "the world economy [does not] concentrate production of each good in a single location" (Krugman, 1998: 8). In fact, economic activity is rarely concentrated entirely in a single location. Instead, economic activities typically fall along a continuum from highly concentrated to widely dispersed. Where do different economic activities fall along this continuum and how could we know?

Few empirical measures of economic geography exist. Measuring geographic patterns of economic activities, such as sectoral employment, is difficult because doing so requires large amounts of highly disaggregate data. Information is needed about where individuals work and what industry or sector they are employed in. All of this information must be available for highly disaggregated geographic units, such as local labor markets. Given these demanding data requirements, many previous measures of economic geography fall short of capturing the theoretical concept of interest. Even more problematically, most existing measures are available only for a single country thereby making cross-country comparisons impossible.

To address these limitations, I generate a continuous empirical measure of the geographic dispersion of economic activity for multiple countries in Chapter 4. With this measure, simplifying assumptions about economic geography are no longer necessary. Instead, I can estimate the actual patterns of geographic concentration and do so for multiple countries in a cross-nationally comparable fashion. Measuring geographic concentration is an important first step in investigating how economic geography influences politics and policy. I leave until Chapter 4 the technical details about the construction of this measure. In the remainder of this chapter, I explore the concept of economic geography: what it is, why does it vary, and how might it matter for politics?

WHAT IS ECONOMIC GEOGRAPHY AND HOW DOES IT VARY?

Economic geography refers to the geographic distribution of economic activities, such as production and employment. I focus primarily on employment because jobs are politically salient. Politicians worry about job losses, and many election campaigns focus on improving job security. Because voters care about jobs, incumbents often go to extraordinary lengths to prevent job losses. In France, for example, the national government placed an order worth €600 million for train equipment it did not need in order to save 480 jobs (Chassany 2016).[2] Jobs matter to voters and consequently the geographic patterns of employment are politically salient.

Employment opportunities are often unevenly distributed between regions within countries. People living in rural Wales, for example, face a different labor market than people living in London. And employment in certain sectors of the economy is unlikely to be distributed evenly between electoral districts or regions of a country (Kendall and Stuart, 1950: 188). In Wales, for example, 11 percent of all jobs were in manufacturing in the first quarter of 2015 (ONS Nomis Database, Workforce Jobs). In contrast, only 2 percent of jobs in London were in manufacturing during the same period (ONS Nomis Database, Workforce Jobs). As this example makes clear, employment opportunities in the manufacturing sector vary across regions within countries.

Different economic sectors exhibit different patterns of geographic concentration. While sectors' exact patterns vary between countries, some sectors are, on average, more concentrated than others. Employment in mining and quarrying tends to be highly concentrated geographically (Shelburne and Bednarzik 1993, Campos 2012). In Norway, for example, a plurality of employees in the oil industry are concentrated in a single electoral district – Rogaland (Statistics Norway 2015). Some service sector activities also exhibit high levels of geographic concentration. Employment in finance and insurance activities,

---

[2] The order included power units for high-speed trains that were not yet able to run on existing rail lines.

information and communication, and professional, scientific, and technical activities tends to be geographically concentrated (Campos 2012). In contrast, other sectors, such as construction, employ people more evenly within a country.

An important distinction between concentrated and diffuse sectors is their need to be located close to their customers. Hairdressers, for example, cannot all be located in one part of the country – they need to be spread out in a pattern similar to the population. Many service providers are geographically dispersed in order to be near end markets, including, for example, hotels and restaurants (Chase 2015). In general, producers of goods or services that serve local customers need to be located close to local populations (Campos 2012). So, whether it is the provision of retail services or restaurants, or the provision of education or social work activities, the distribution of jobs in these industries is generally spread across countries in order to reflect their need to directly serve local populations (Campos 2012). Industries that do not need such close location to customers – such as car manufacturing plants – are often more geographically concentrated (Campos 2012). For example, motion picture production and investment banking are each geographically concentrated precisely because they do not need to be located close to their end customers (Kolko 2010).

Agriculture tends to be concentrated in areas where there are few alternative employment opportunities. In other words, agriculture exhibits high levels of *relative* concentration – that is, it is highly concentrated relative to the distribution of aggregate economic activity (Shelburne and Bednarzik 1993, Brülhart and Traeger 2005). In many rural communities in the Great Plains of the United States, for example, nearly everyone is either directly or indirectly employed in the agriculture sector.

The manufacturing sector on average tends to be less concentrated than agriculture. However, the geographic patterns of manufacturing sector employment vary notably between countries. Among OECD countries, manufacturing employment is the most geographically concentrated in Sweden and Australia (OECD 2007). It is the least geographically concentrated in the Czech Republic and Denmark (OECD 2007). Manufacturing employment in Sweden is 3.5 times more geographically concentrated than manufacturing employment in Denmark (OECD 2007). In the United States, manufacturing employment is nearly 1.5 times more concentrated than manufacturing employment in the United Kingdom (OECD 2007). As these data make clear, significant variation exists in the geographic dispersion of manufacturing employment between developed countries.

WHY DOES ECONOMIC GEOGRAPHY VARY?

The geographic concentration of sector employment varies within and between countries. An obvious question is why. What explains the variation in economic geography? A large literature addresses this question, and many theories seek to explain why the geographic concentration of economic activity occurs (Marshall 1920, Krugman 1991). As discussed above, a key distinction between concentrated and diffuse industries is their need to be located close to their customers. Various other factors also differentiate concentrated sectors from diffuse ones. The mobility of capital and labor can help to explain the uneven distribution of economic activity (Maloney and Nayyar 2017). Mobility refers to the costs of moving factors of production (i.e. land, labor, and capital) to a new use in the domestic economy (Hiscox 2002, Rickard 2009). These costs may systematically influence the geographic distribution of economic activities. Factors facing high adjustment costs may decide not to move. As a result, high adjustment costs reinforce existing patterns of economic geography. In contrast, factors facing low adjustment costs can move with relative ease and are therefore more likely to relocate to take advantage of attractive features (Maloney and Nayyar 2017).

In addition to adjustment costs, other factors may also influence producers' (re)location decisions. Larger markets offer better matching between employers and employees, buyers and suppliers, and entrepreneurs and financiers (Maloney and Nayyar 2017). To take advantage of better matching opportunities, producers may move to larger markets, which could lead to ever greater geographic concentration and contribute to urban agglomeration. Producers may also cluster together to learn about new technologies, market evolutions, or new forms of organization (see Krugman 1991, Marshall 1920). Additional benefits of concentration include transportation costs and intellectual spillovers (Dumais, Ellison, and Glaeser 2002). In sum, the origins of economic geography are complicated and wide-ranging.

Throughout this book, I take economic geography as given. In other words, I do not attempt to explain the geographic patterns of employment. Instead, I investigate the political implications of existing patterns of economic geography. This strategy makes sense given the focus of my argument and is reasonable given that patterns of economic geography remain relatively stable over the medium term (Dumais et al. 2002). Furthermore, patterns of economic geography appear to be largely exogenous to politics and policy.

Geographic patterns of economic activity appear to be largely immune to government subsidies. The US biotechnology industry, for example, is concentrated in five urban centers. This high level of geographic concentration persists despite the fact that forty-one out of fifty US states have significant funding programs to spur development of the life sciences industry (OECD 2007). As the US biotech example suggests, firms' location decisions are often unresponsive to government subsidies (Midelfart-Knarvik, Helene, and Overman 2002). Quantitative statistical tests show the effect of government subsidies on firms' location decisions is negligible. In the United Kingdom, an increase in the expected government subsidy of £100,000 is associated with only a 1 percent increase in the probability of (re)location (Devereux, Griffith, and Simpson 2007). Government subsidies do not appear to significantly affect firms' location decisions (Devereux et al. 2007). Furthermore, civil servants responsible for the allocation of government-funded subsidies report that they have never seen a firm relocate to try to win more subsidies.[3] Bureaucrats I interviewed in Norway acknowledged that it was more difficult for a firm in Oslo to win subsidies because of the higher concentration of firms in that district. Norway uses proportional electoral rules so this observation is fully consistent with my argument: geographic concentration handicaps groups seeking assistance from governments in PR systems. However, the bureaucrats stated that no firm had ever moved out of Oslo to increase their chances of winning subsidies from the government.[4]

Similarly, electoral institutions do not appear to significantly influence economic geography. If actors anticipated the benefits of geographic concentration in plurality systems, they would cluster together with others to win more generous subsidies. But the geographic concentration of manufacturing employment is no higher in plurality systems than in PR systems, on average.[5] Among OECD countries, those with proportional electoral systems exhibit both the highest (Sweden) and lowest (Denmark) levels of manufacturing-sector geographic concentration (OECD 2007). The highest values of concentration are observed in a PR country (Sweden) and a plurality country (Australia), and these values are similar in magnitude (54 and 51 respectively) despite the countries' different electoral systems (OECD 2007). These observations suggest that producers' location decisions are not driven by a country's electoral institutions and help to minimize concerns about causal complexity.

---

[3] In-person interview with Innovation Norway staff member Pål Aslak Hungnes and Per Melchior Koch in Oslo, Norway on June 19, 2014.
[4] In-person interview with Innovation Norway staff members Pål Aslak Hungnes and Per Melchior Koch in Oslo, Norway on June 19, 2014.
[5] Author's calculations.

Geographic concentration can occur "naturally" through market mechanisms (Campos 2012, Maloney and Nayyar 2017).[6] The pull factors that stimulate natural clusters to arise include the availability of raw materials, suitable climate conditions, and proximity to markets (Maloney and Nayyar 2017). Coordination on certain locations can also be achieved by the strategic actions of large players, such as universities and multinational corporations, or private groups such as export business associations, credit cooperatives, and industry associations (Maloney and Nayyar 2017). In other words, economic geography is shaped by numerous factors – many of which have little, if anything, to do with government policy.

THE GEOGRAPHY OF NARROW INTERESTS

The varied geographic distribution of sector employment illustrates an important point: special interests may or may not be geographically concentrated. Consider for example the Swedish automotive industry. It consists of three key producers: Volvo, Saab, and Scania AB. Together, these three firms employ only a small fraction of Sweden's total population. In this way, the Swedish automotive industry represents a special or "narrow" interest. *Narrow interests* are defined as a group with a common economic interest shared by a small segment of the population. In this example, the narrow interest happens to be geographically concentrated. Saab employed 4,100 people in 2009 and 3,700 of these employees, or 90 percent of the firm's total workforce, worked at its hub in the southwestern city of Trollhättan. Saab was, in fact, the largest employer in Trollhättan. The second largest employer was the municipal government. This example illustrates a geographically concentrated narrow interest.

Not all narrow interests are geographically concentrated. In fact, some are quite diffuse. In the United States, for example, the cosmetic sales industry is dominated by a single firm: Mary Kay. In 2013, Mary Kay employed approximated 5,000 people and these employees were spread out across the entire country.[7] Indeed, nearly every community had at least one Mary Kay employee. This example illustrates a geographically diffuse narrow interest. Such interests can and do win government support, particularly in PR countries. The Austrian government, for example, provided subsidies to fund the purchase of microwave ovens

---

[6] Given certain assumptions, most notably an economy displaying increasing returns to scale and monopolistic competition.
[7] www.forbes.com/companies/mary-kay/.

by restaurants (Sturn 2010). This subsidy was targeted to a narrow interest group (i.e. restaurant owners) but one that was geographically diffuse. Restaurant owners across the entire country benefited from the government-funded subsidy.

As these examples illustrate, geographic concentration is conceptually distinct from the notion of narrow (or "special") interests. Scholars often conflate these two concepts by assuming that all special interests are geographically concentrated. This incorrect assumption generates confusion about when and under what conditions special interests have the most political power. Pulling apart the distinct concepts of "narrowness" and geographic concentration can help to untangle the politics of particularistic economic policies.

In this book, I use interchangeably the terms narrow interests and special interests. Both of these terms refer to a common economic interest shared by a small segment of the population. The characteristic that makes an interest "narrow" is the portion of a country's population that shares the same economic interest. The US steel industry, for example, makes up a narrow interest group because it employs just 0.3 percent of the total US population. People employed in the steel industry share a common economic interest in policies like tariffs that protect the industry from lower-cost imports from China. This narrow interest is geographically concentrated in three of the fifty US states.

Economic geography refers to the dispersion of economic activity across space. More precisely, economic geography refers to the geographic dispersion of employment in a given sector or industry. People employed in the same industry often share common economic interests, most notably in the economic fortune of their industry (Hiscox 2002). The geographic pattern of industrial employment can therefore be used to estimate the geographic concentration or diffusion of narrow economic interests. Using these definitions, I separate the concepts of narrowness and geographic concentration. Once these two concepts are disentangled from one another, new insights emerge regarding politics and policy. However, before moving on to my argument, I first briefly discuss how economic geography may influence politics.

ECONOMIC GEOGRAPHY AND POLITICS

Growing evidence highlights the political importance of economic geography. Recent studies demonstrate that economic geography mediates the impacts of globalization. Rising Chinese imports have been shown to have varied effects on different communities because of the uneven geographic distribution of manufacturing employment within countries (Autor et al. 2013, Ballard-Rosa et al. unpublished manuscript). Some regions experience significant reductions in manufacturing employment and wages, while others see few negative effects from the rising tide of Chinese goods. Because of the uneven geographic distribution of economic activities, the losers from globalization tend to be concentrated in certain locations and falling mobility means that economic hits are not shared between regions as much as before. The costs of globalization therefore fall particularly hard on some local communities because of economic geography.

Globalization's uneven economic effects have important political consequences. In the United States, for example, voters in congressional districts exposed to larger increases in foreign imports disproportionately removed moderate politicians from office in the 2000s (Autor et al. 2016). This evidence suggests that economic geography contributes to increasing political polarization. Although economic geography likely has important consequences for politics, many – but not all – studies have overlooked the geographic distribution of economic activity to date, as I discuss below.

District Size

Economic geography has played an important – although implicit – role in theories about the consequences of district size. District size refers to the number of people living in an electoral district. An electoral district is a geographically defined area within which votes are counted and seats allocated (Cox 1997). Several scholars speculate that larger districts will better insulate politicians from protectionist demands while smaller districts will give greater influence to parochial interest (e.g. Rogowski 1987, Alt and Gilligan 1994, Mansfield and Busch 1995, McGillivray 2004). The logic of this argument relies on economic geography. McGillivray (2004: 28) provides an illustrative example:

"An industry with 100 employees represents 10 percent of the electorate in a district with 1,000 voters. The same industry represents only 0.1 percent of the electorate in a district of 100,000 voters. In the larger district, refusing to protect the industry is unlikely to affect a politician's reelection chances in a plurality system because the industry is only 0.1 percent of the representative's electorate."

As this illustration makes clear, it is impossible to hypothesize about the effects of district size on policy "without making further assumptions about the [geographic] distribution of industries" (Hatfield and Hauk, 2014: 522). In fact, most arguments about district size make implicit assumptions about the geographic distribution of economic activities. Yet, few theories explicitly discuss economic geography and even fewer attempt to measure it empirically.

Many empirical studies of the policy effects of district size make cross-national comparisons. Rather than measuring district size directly, these studies instead assume that proportional systems have larger districts than plurality systems. They then compare policy outcomes in plurality countries with those in proportional countries and assert that the observed policy differences are due to district size.

Rather than assuming that PR systems have larger districts, Hankla (2006) instead constructs a direct measure of district size. He divides the total number of seats in a country's lower legislative chamber by mean district magnitude and divides that number into the country's total population to estimate the number of people living in an average district in a given country. Using this measure of district size, Hankla (2006) finds that countries with larger districts tend to have lower levels of trade protection, on average. Similarly, Rogowski (1987) reports evidence that suggests larger districts are associated with lower trade barriers. He finds a negative correlation between the number of parliamentary constituencies and trade openness and asserts that the number of constituencies is an inverse measure of average constituency size.

Although many empirical investigations of district size draw comparisons between countries, a series of studies exploit the differences in constituency size within a single country – most often the United States. Results from these intra-country studies are mixed. Some suggest that industries in smaller constituencies receive more trade protection than industries in larger constituencies (Hauk 2011). Others find no relationship between district size and trade protection (Karol 2007, Hatfield and Hauk 2014). For example, constituency size does not seem to account for the differences in trade policy preferences among the House, Senate, and presidency (Karol 2007). In fact, the positive correlation between plurality systems and trade protection remains robust to the inclusion of measures of district size (Hatfield and Hauk 2014). This evidence rules out district size as the causal mechanism linking electoral systems to policy outcomes.

Plurality Electoral Systems

Although most studies of politics overlook economic geography, some explicitly examine the political consequences of geographic concentration. Yet virtually all of these studies focus exclusively on plurality systems. McGillivray examined the effects of geographic concentration in two plurality countries: the United States and Canada (McGillivray, 1997: 2004). In these countries, McGillivray found that concentrated industries tend to win more trade protection than diffuse industries. However, McGillivray does not test the effects of geographic concentration in PR countries. Likewise, Hansen (1990) examines the political effects of concentration only in the United States. She established that geographically concentrated industries are more likely to secure favorable rulings for antidumping claims. Similarly, Milner (1997) showed that concentrated industries in the United States made fewer concessions in negotiations over the North American Free Trade Agreement (NAFTA). In sum, previous empirical studies report a positive correlation between geographic concentration and political influence in plurality systems.

The political consequences of geographic concentration remain unknown in proportional electoral systems. Does geographic concentration confer similar benefits to groups in PR systems as in plurality systems? Busch and Reinhardt (2005) provide some evidence to suggest concentration may be politically valuable in PR systems. They report that citizens employed in more geographically concentrated industries in the Netherlands – a PR country with a single national district – are more likely to turnout to vote. However, their empirical measure of turnout is, in fact, a measure of party vote intention. Their results are therefore more correctly interpreted as evidence that citizens employed in concentrated industries are more likely to express a partisan preference. In other words, citizens employed in geographically concentrated industries may be less likely to be swing voters.

With the notable exception of Busch and Reinhardt (2005), virtually all research on the political effects of geographic concentration focuses on plurality systems. This fact presents a challenging inference problem. Existing studies examine only a subset of democracies, and the chosen subset is not random; it consists only of democracies with a specific set of electoral institutions, namely single-member districts and plurality electoral rules. It is in precisely such systems that geographic concentration is most likely to matter. Where politicians compete for office in geographically defined electoral districts, the concentration of economic interests will be important for their political influence. In other words, previous studies look for and find evidence that geography matters precisely where geography is most likely to matter – in plurality systems with geographically based electoral competition.

Existing research tell us little about how geography matters in countries with different electoral institutions. In proportional systems, for example, legislative seats are awarded to parties based on their vote shares.

As a result, all votes may be equally valuable to parties competing in PR systems regardless of their geographic location. Given this, it is unclear what, if any, effect geography might have on politics and policy-making in PR systems. Cognizant of this issue, McGillivray recommended that future research investigate the effects of geographic concentration in proportional rule systems (McGillivray, 1997: 604). My research represents an effort to respond to this appeal by investigating the effects of geographic concentration in various electoral systems.

3

How Institutions and Geography Work Together to Shape Policy

Electoral institutions and economic geography work together to shape economic policy. Economic policy is agreed by incumbent politicians and government parties whose decisions are often influenced by a desire to win reelection. Incumbents' optimal reelection strategy depends on a country's electoral institutions. Economic geography defines the best means for incumbents to achieve their institutionally generated reelection strategy.

Economic geography and electoral institutions align at times to produce incentives for leaders to supply certain types of economic policies. In this chapter, I identify the conditions under which leaders face the greatest incentives to supply particularistic economic policies, such as subsidies. I discuss two mechanisms through which economic geography and electoral institutions work together to shape leaders' incentives: (1) effective vote maximization and (2) the nature of electoral competition.

Parties and politicians competing in democratic elections seek to maximize their effective votes – that is the votes needed to win office. To maximize their effective votes and consequently their chances of reelection, parties and politicians strategically employ economic policies. The efficacy of a particular economic policy for "effective vote maximization" depends on the geographic distribution of its beneficiaries. Policies that benefit geographically concentrated groups provide the best opportunity for politicians to maximize their effective votes in plurality systems. As a result, programs for geographically concentrated groups tend to be especially generous in plurality countries. In the United States, for example, where politicians win office by securing a plurality of votes in single-member districts, concentrated industries habitually win generous subsidies. US sugar cane farmers receive an extra $369 million dollars a year from government-funded subsidies and nearly 60 percent of this money goes to just 17 growers in a single state: Florida (Beghin et al. 2003, Frieden et al. 2010). Similarly, cotton producers concentrated in a single electoral district (the 19th district of Texas) won $180 million dollars in subsidies from the government in 2014. Geographically concentrated producers of footwear and automobiles also receive generous government assistance in the United States.

Incumbents fund subsidies for business to win votes (Dewatripont and Seabright 2006, Buts, Jegers, and Jottier 2012). Subsidies engender economic benefits, which often garner votes for incumbents. Manufacturing subsidies, for example, benefit citizens who own production factors, such as labor or capital, used in manufacturing industries (Persson and Tabellini, 2003: 14). Subsidies raise producers' income beyond that which would be earned without intervention (Schwartz and Clements 1999). Subsidies can also increase wages to levels above those available in a purely competitive market (Mortensen and Pissarides 2001). Subsidies help to protect the industry from competition with lower-cost foreign imports and maintain wages and employment.[1]

Subsidies have also been used to persuade producers to keep facilities open and to retain jobs which in other circumstances would not be viable (Sharp 2003). Subsidies to the British textile industry, for example, were funded by the government for precisely these reasons (Duchêne and Shepherd 1987, McGillivray 2004). Following a rapid increase in competition from foreign imports, the British government provided generous subsidies to the geographically concentrated textile industry in an attempt to maintain employment and wage levels in the industry (Duchêne and Shepherd 1987). More generally, subsidies can reduce the negative impacts of disturbances to the economy, economic downturns, or changes in trends. Subsidies may be necessary to help businesses survive, preserve jobs, and prevent the loss of skills. Subsidies can also help to "ensure the security of particular supply chain considered essential for the functioning of modern industrialized economy" (OFT 2009).[2] Sector-specific subsidies can also serve to improve export performance (Aghion et al. 2011, Stöllinger and Holzner 2016) and increase employment and investment (Criscuolo et al. 2012). In short, subsidies engender economic benefits, which may garner votes for incumbent politicians and parties.

---

[1] Other similar programs may include tax incentives, relief from industry-specific regulations, and industry-specific trade barriers. See, for example, Singer (2004).
[2] The Office of Fair Trading (OFT) was responsible for protecting consumer interests throughout the UK. It closed on April 1, 2014, with its responsibilities passing to a number of different organizations including the Competition and Markets Authority (CMA) and the Financial Conduct Authority.

Although subsidies engender economic benefits, they are not equally efficient at garnering votes in all circumstances. The efficacy of subsidies as a vote-winning policy tool depends on a country's electoral institutions and economic geography. In PR systems, parties maximize their effective votes by assisting geographically diffuse groups. Subsidies to diffuse groups enable parties to compete for votes in large areas of the country. In PR systems with multiple electoral districts, winning votes across the entire country helps parties maximize the number of seats they hold in parliament. Although focusing exclusively on geographically concentrated groups may increase a party's vote share in a given district, it limits the national appeal of a party and consequently the number of legislative seats a party could win. In Norway, for example, a party called People's Action Future for Finnmark (*Folkeaksjonen Framtid for Finnmark*) focused entirely on improving economic conditions for the fishing industry in Finnmark and subsequently won 21.5 percent of the vote in the district in 1989 (Aardal 2011). However, the party won just 0.3 percent of the national vote.

Seeking assistant for geographically concentrated groups, like Finnmark's fishing industry, can limit the national appeal of parties and consequently their share of the national vote. As a result, parties that pursue geographically targeted programs may fail to cross national vote-share thresholds, which are required in many PR countries. These thresholds stipulate that political parties must win a minimum share of the national vote in order to hold any seats in parliament. In Slovenia, for example, the threshold is 4 percent, as required by the National Assembly Elections Act of 2000. Legislative seats are apportioned only to political parties that reach the 4 percent threshold. Similar thresholds exist in other PR countries; in Poland, Germany and New Zealand, for example, the threshold is 5 percent and in Sweden, there is a nationwide threshold of 4 percent. Because of the national vote-share thresholds that exist in many PR countries, parties cannot afford to focus their attention on geographically concentrated groups because doing so may effectively exclude them from parliament.

By focusing on geographically concentrated groups, parties fail to maximize the number of legislative seats they could win in a PR system. For example, the Norwegian party People's Action Future for Finnmark (*Folkeaksjonen Framtid for Finnmark*) demanded increased government assistance for the fishing industry, which was concentrated largely in the northeastern region of Finnmark (Aardal 2011). The party won 21.5 percent of the vote in the district of Finnmark in 1989 and as a result, the party won one of the district's legislative seats (Aardal 2011). However, the party was not eligible for any of the seats allocated at the national level (i.e. compensatory seats) because it failed to clear the national threshold of 4 percent. As this example illustrates, parties competing in PR systems have powerful incentives to pursue votes across the country. To maximize their effective votes and consequently the number of legislative seats they hold, parties in PR systems work to assist geographically diffuse groups.

Given these incentives, I hypothesize that government-funded programs will be relatively more generous in PR countries when the potential beneficiaries are geographically diffuse. The geographically diffuse forestry sector in Sweden, for example, receives generous subsidies. In Sweden, where elections are held via proportional representation, the forestry industry won 10 percent of all government subsidies in the late 1970s. In fact, the government spent nearly twice as much on subsidies to the geographically diffuse forest sector as it did on subsidies to the geographically concentrated textile industry (Carlsson, 1983: 11). This pattern suggests that geographic concentration may be something of a political liability in PR countries. The Swedish government has, in fact, refused to provide financial assistance to geographically concentrated groups. For example, the government refused to assist the geographically concentrated automotive manufacturer Saab. Ninety-percent of Saab's employees worked in the southwestern town of Trollhättan. It was virtually impossible to find anyone in Trollhättan who was not somehow connected to Saab. Saab was the largest employer – providing work for even more people than the municipal government – and Saab's economic difficulties spelled potential disaster for the city. Yet, the Swedish Prime Minister, Fredrik Reinfeldt, refused to bail out Saab, saying he would not put "taxpayer money intended for healthcare or education into owning car companies" (Ward 2009). The Swedish Enterprise and Energy Minister Maud Olofsson told Swedish public radio that "voters picked me because they wanted nursery schools, police and nurses, and not to buy loss-making car factories."[3] This anecdotal evidence suggests that leaders in proportional systems, like Sweden, are less responsive to geographically concentrated groups, as compared to more geographically diffuse groups.

In the following section, I discuss two mechanisms through which economic geography and electoral institutions shape leaders' incentives and subsequently policy: (1) effective vote maximization and (2) the nature of electoral competition. I then discuss other possible mechanisms that might link institutions and geography to economic policy.

[3] http://news.bbc.co.uk/1/hi/business/7899244.stm.

MAXIMIZING EFFECTIVE VOTES

Votes can be split into three categories: "wasted votes" that play no part in winning a legislative seat; "effective votes" – those votes needed to win; and "surplus votes" – those votes that are not needed for victory (Johnston, Rossiter, and Pattie 2006). In plurality systems, defining the number of effective votes is simple – it equals one more vote than the second-place candidate's total. However, the definition of effective votes varies between electoral systems. Votes that are "surplus" in a plurality system can be "effective" votes in a proportional system. Imagine, for example, a candidate wins 60 percent of a district's votes. In a plurality system with single-member districts, the candidate could have won the seat with many fewer votes. By winning 60 percent of the district's vote, the candidate earned many "surplus" votes. Surplus votes are wasted in a plurality system – that is they play no part in winning office and do nothing to help other candidates from the same party. In contrast, surplus votes go to help candidates from the same party in a PR system with multimember districts. In other words, surplus votes in plurality systems are effective votes in proportional systems. Given a country's institutional framework, parties and politicians seek to maximize their effective votes.

Parties and politicians can use economic policies, like subsidies, to maximize their effective votes. Subsidies win votes. Incumbents who spend more on subsidies win more votes in subsequent elections (Buts et al. 2012).[4] However, the efficacy of subsidies as a vote-winning policy tool depends on a country's electoral institutions and the geographic concentration of the subsidy's beneficiaries. To illustrate this point, consider the following example. Imagine a sector of the economy that employs 10 percent of the voting-age population. (The tourism industry in the United Kingdom is roughly this size; it accounted for 9.6 percent of total employment in 2013.)[5] An incumbent party could potentially secure (or increase their chances of securing) 10 percent of the national vote by subsidizing this sector.[6] Yet depending on the geographic pattern of employment in the sector, subsidies might buy an additional 3 percent of the vote in one district and 30 percent more of the vote in

---

[4] Levitt and Snyder (1997) estimate that it takes about $14,000 US dollars in federal funding to win an additional vote. In Australia, Leigh (2008) finds a strong relationship between government spending on the Roads to Recovery Program and increased voter support for incumbents.
[5] www.tourismalliance.com/downloads/TA_369_395.pdf.
[6] Opposition parties could potentially secure 10 percent of the national vote by credibly promising to increase subsidies to the sector once in office.

another. In other words, the potential electoral gains from subsidies depend on the geographic distribution of the subsidy's beneficiaries.

Sectors with geographically diffuse employees can, in exchange for subsidies, yield votes across the country. In contrast, sectors with geographically concentrated employees can supply votes in only select areas. The potential value of these two options for incumbents seeking re-election depends on a country's electoral institutions.

In Proportional Systems

In a PR system with one nationwide electoral district, the geographic distribution of a sector's employees is politically irrelevant. All votes contribute equally to the number of legislative seats won by the party, regardless of their geographic location. In a fully proportional single-district system, a 10 percent gain in a party's share of the national vote will translate into 10 percent more legislative seats for that party. Parties competing in PR systems with a single, nationwide electoral district therefore work to maximize their share of the national vote.

The story is more complicated in PR systems with multiple, subnational electoral districts. In PR systems with multiple districts, geography matters. Most legislative seats are awarded to parties based on their district-level vote shares – rather than their national vote share. In Norway, for example, 150 of the 169 legislative seats are distributed at the district level to parties in proportion to their share of the district-level votes. Given this, the geographic location of potential voters is important to parties competing in PR systems with multiple subnational electoral districts.

Electoral gains can be uneven across a country's electoral districts. Imagine a party gains an average of 10 percent of the vote from one election to the next. There are two different ways in which this gain could be distributed across electoral districts. First, the party could gain exactly 10 percent of the vote in all districts. This outcome represents a "uniform" vote swing (i.e. a gain of 10 percent in all districts). Alternatively, the party could gain more than 10 percent in some districts and less than 10 percent in others, and this "uneven" vote swing yields an average swing of 10 percent over all districts (Tufte 1973).

A uniform swing generates different election results than an uneven swing. In PR countries with multiple electoral districts, legislative seats are awarded to parties based on their district-level vote shares – rather than their national vote share. In a PR system with multiple electoral districts, a uniform vote swing typically yields more seats for parties than an unevenly distributed swing (Tufte 1973). Given this, parties work to achieve a relatively uniform vote swing and to this end subsidize geographically diffuse groups. Assisting sectors whose employees are spread across the country produces a more uniform swing in the party's vote share than subsidizing geographically concentrated sectors. Given this, I hypothesize that parties competing in PR systems with multiple electoral districts will spend relatively more on subsidies for geographically diffuse groups.

Geographically diffuse industries tend to win generous subsidies in PR systems. Even in mixed-member proportional systems, where some legislators are elected from single-member districts and others are elected from a party list, geographic diffusion appears to confer an advantage on interest groups. In Germany, for example, where legislators are elected via a mixed-member system and each party's share of seats in parliament is proportional to the number of "second votes" it received,[7] the government subsidizes geographically diffuse sectors. The German forest sector is geographically diffuse; employment in the sector is roughly similar to the overall distribution of the German population (Kies, Mrosek, and Schulte 2009). People work in the forest sector in every region of Germany (Kies et al. 2009). The sector accounts for 5 percent of total employment in 19 percent of the counties in Germany and in 30 countries it accounts for 7.5 percent of total employment (Kies et al, 2009: 44). This geographically diffuse sector receives generous subsidies from the government. Thirty percent of government-funded subsidies go to the forestry sector (Thöne and Dobroschke 2008). The total volume of these subsidies amounted to €10.8 billion in 2006 (Thöne and Dobroschke 2008). The geographic diffusion of the German forestry sector helps to explain why it wins generous government subsidies – even though it employs only a small share of Germany's total population. Although the sector employed just nine hundred thousand people in 2004 (Kies et al., 2009: 39), its employees were spread across the country and consequently the sector was electorally valuable to parties competing for office in a mixed-member proportional system. Because legislative seats are awarded to parties in proportion to their share of second votes, government parties have incentives to provide subsidies to geographically diffuse groups and, in fact, many Germany governments have routinely subsidized diffuse groups, such as the forestry industry (Schatz and Wolter 1987, Thöne and Dobroschke 2008).

---

[7] In German elections, every voter gets two votes. The first allows voters to choose an individual candidate to represent their district. The "second vote" is for the party they support.

In Plurality Systems

Although district-specific gains are politically relevant in all countries with multiple electoral districts (Tufte 1973), district-specific gains are more electorally salient in plurality systems than PR systems (Tufte 1973). Winning more than a plurality of votes in any given district is futile because the surplus votes are wasted in a plurality system – that is they do not help the candidate or the candidate's party. Winning 60 percent of the vote, for example, is overkill. The candidate would have won the election with many fewer votes and the surplus votes are wasted. Given this, political parties competing in plurality systems seek to avoid "over buying" support in a given district. In plurality systems, parties aim to minimize surplus vote while maximizing effective votes.

Assisting groups spread across a country will likely "over buy" support in some districts. In contrast, targeting benefits to key districts is more efficient for political parties competing in plurality systems. When beneficiaries are geographically concentrated, politicians and parties can use subsidies to target benefits to vital districts to ensure the maximum number of effective votes. Given this, I hypothesize that geographically concentrated groups will receive more generously funded subsidies in plurality systems than PR systems and concentrated groups will win relatively more subsidies than diffuse groups in plurality systems, all else equal.

France's two-ballot majority-plurality system, for example, incentivizes the provision of subsidies to geographically concentrated producers. As a result, the concentrated audiovisual services industry in France receives generous state support (Creative Screen Associates 2013). The industry is highly concentrated – located entirely in and around the Île-de-France region (Dale 2015). The industry's geographic concentration contributes to its political clout. For example, the industry's geography helped it win protectionist amendments to the 1946 Franco-American trade agreement, also known as the Blum–Byrnes agreements (Frey 2014). The industry took to the streets of Paris to protest the Franco-American trade agreement. Tens of thousands of people marched through central Paris shouting *"Cinéma français."* The demonstrations were organized by a committee, which met at the national film school IDHEC in Paris, established to defend cinema from US imports. The film industry ultimately prevailed when the National Assembly revised the agreement to grant protection and financial assistance to the industry (Frey 2014).

The film industry's geographic concentration in Paris helped it win protectionist amendments to the 1946 Franco-American trade agreement. Politicians from the region were beholden to the industry for electoral support and consequently worked to assist the industry. French politicians continue to do so today; France's film industry is subsidized to the tune of about €1 billion a year (Carnegy 2013).

In contrast, the geographically diffuse film-related technical industry receives little government assistance.[8] French camera equipment manufacturers, for example, receive far less state support than other segments of France's audiovisual industry (Dale 2015). Camera equipment producers lack the political influence needed to win generous subsidies in France's majority-plurality system because of their geographical dispersion. The two largest manufacturers of motion picture equipment are located at nearly opposite ends of France – Aaton is based in Grenoble and Transvideo is based in Normandy. Industry leaders bemoan this geographic dispersion. "In France we don't have a geographical concentration of equipment manufacturers, as you find in places such as L.A., London or Munich," explains a leading figure in the French film-equipment industry.[9]

The industry's geographic dispersion limits politicians' incentives to subsidize film-equipment manufacturers. The electoral rewards from supplying subsidies to camera equipment manufacturers cannot be captured by any single legislator because of the industry's geographic diffusion. Efforts by a legislator from Grenoble to assist the industry will benefit legislators from Normandy who may not be from the same political party or ideological persuasion. Geography thus helps to explain why the diffuse technical industry receives so little government assistance, especially when compared to the far more concentrated film production industry.

In an attempt to increase their political power, France's leading film equipment manufacturers launched a new trade organization in March 2014. The association's main goal is to lobby the French government for subsidies and other benefits. "We need to make the French government realize that it's a real industry, with hundreds of employees and a high level of creative and technological innovation," states Delacoux, the president of the new organization (Dale 2015). This trade organization demonstrates that geographically diffuse industries

---

[8] The variation in state support within the audiovisual industry is particularly interesting given France's claim that the industry is "culturally sensitive." France often seeks exceptions to international economic agreements for the film industry, arguing that it requires special treatment because of its cultural importance. Yet, the entire industry does not receive equal levels of state support. French camera equipment manufacturers receive far less aid than other segments of France's audio-visual chain.

[9] Delacoux quoted in Dale (2015).

can and do come together to form interest groups. Conventional wisdom posits that geographically diffuse industries face higher collective action costs than concentrated industries. While this may be true, the relatively higher collective action costs are not prohibitive – that is, they do not prevent collective action in all cases, as illustrated by the French film equipment manufacturers.

THE NATURE OF ELECTORAL COMPETITION

In addition to the vote-swing mechanism, a second mechanism links electoral institutions and economic geography to policy, namely the nature of electoral competition. Electoral competition can be characterized as being either candidate-centered or party-centered. Party-centered competition encourages voters to emphasize their party preference over that for specific candidates (Carey and Shugart 1995). Voters decide which party to support based on the parties' platforms and election promises – rather than any individual candidate's personal characteristics. In contrast, candidate-centered competition encourages voters to see the basic unit of representation as the candidate rather than the party (Shugart, 1999: 70).

Candidate-centered electoral competition often emerges in countries with plurality electoral rules and single-member districts (Carey and Shugart 1995). In such systems, voters decide which candidate they want to represent them in the legislature. Their decision depends primarily on a candidate's individual characteristics and/or record rather than the party to which the candidate belongs (Carey and Shugart 1995). To maximize their chances of (re)election in such systems, politicians work to develop a personal vote. A personal vote is that part of a legislator's electoral support that is based their personal reputation. Generating a personal vote maximizes candidates' chances of winning office when elections are candidate-centered (Fenno 1978, Weingast, Shepsle, and Johnson 1981, Lancaster 1986, Lancaster and Patterson 1990, Cox and Rosenbluth 1993, Carey and Shugart 1995).

France's majority-plurality system generates incentives for politicians to cultivate a personal vote. A survey of sitting French Members of Parliament (MPs) revealed that 41.2 percent agreed that it was their duty to "represent above all his/her constituency and his/her region" (Brouard et al. 2013).[10] As one French MP put it, "[a]n MP is a gardener. He has a big garden, his constituency, and he has to go to

---

[10] Brouard et al. (2013) find that legislators work harder for their constituents when they feel that they alone are responsible for their reelection chances.

Paris in order to get fertiliser."[11] Subsidies are one example of such "fertilizer." When producers are geographically concentrated in a MP's district, legislators have powerful electoral incentives to secure subsidies for producers in their districts because doing so helps politicians develop a personal vote.

The usefulness of subsidies for building a personal vote depends on economic geography. Economic geography refers to the geographic pattern of employment for a given sector or industry in a given country. Imagine, for example, that an industry's employees are located entirely within a single electoral district. A legislator can increase their chances of reelection by providing a subsidy to that industry in a candidate-centered system. The benefits of such a subsidy will be concentrated in the politician's electoral district and the beneficiaries have the opportunity to vote for the legislator. In this way, subsidies to geographically concentrated industries are an efficient means for incumbents to maximize their chances of reelection in candidate-centered systems.

Politicians whose electoral success depends on support only from their geographically defined constituents have few incentives to cater to geographically dispersed interests. Doing so neither sufficiently rewards their efforts nor maximizes incumbents' chances of reelection because the electoral rewards for catering to diffuse interests are spread over many districts. In an industry where employment spans an entire country, the benefits of industry-specific subsidies accrue to voters in all electoral districts. In this case, promoting subsidies would be an inefficient way for a politician to "buy" the votes she needs to win in a candidate-centered system.

The geographic concentration of citizens with shared economic interests is relatively less important in party-centered systems. Parties, not individual legislators, are the primary actors in party-based systems. Political parties define both electoral strategies and public policy (Tavits 2009). Parties seek to win as many effective votes as possible to maximize their parliamentary representation. Thus, parties have an incentive to appeal to a diffuse constituency that "displaces the district as the primary electoral constituency" (Lancaster and Patterson, 1990: 470).[12] Appealing to a diffuse constituency may engender "a shift in the national mood towards the ruling party" (Reed et al. 2012) and such a shift may increase the party's vote share across the country.

---

[11] Quoted in Brouard et al. (2013: 146). This quote came from a member of the centre-right political party Union for a Popular Movement (*Union pour un mouvement populaire*/UMP).

[12] "Thereby decreasing the importance of pork-barrel politics' (Lancaster and Patterson, 1990: 470).

In proportional systems, legislative seats are awarded in accordance with parties' vote shares. Because every vote contributes to the allocation of legislative seats among parties, politicians and parties competing in proportional systems have incentives to cater to citizens' economic interests, even when they are geographically diffuse. Winning the support of voters in an industry that employs only 1 percent of the population could, for example, translate into several additional legislative seats. Furthermore, winning an additional 1 percent of the national vote may mean the difference between being in or out of parliament, because many proportional systems have minimum threshold requirements that require a party receive a minimum percentage of votes to obtain any seats in the parliament.

This logic helps to explains why geographically diffuse groups win economic benefits in party-centered, PR systems. But why would governments elected in party-centered PR systems provide relatively fewer subsidies to concentrated groups than diffuse groups? The answer is simple: subsidies entail costs. These costs include both real monetary costs and opportunity costs. Every dollar spent on subsides for concentrated groups is one less dollar available for subsidies to diffuse groups. Subsidies for diffuse groups bring relatively greater electoral benefits to parties competing for office in party-centered PR systems, so parties prioritize spending on diffuse groups.

Subsidizing a geographically diffuse sector helps people across the country. Subsidizing the geographically diffuse construction industry, for example, helps citizens across virtually all regions. Employees in the construction industry benefit directly from subsidies via increased wages and more secure employment. Owners of capital invested in the industry also benefit from above-market rates of return and greater demand. Subsidies to the construction industry also benefit related sectors, such as real estate and retail. In this way, subsidies to one industry can indirectly help other related industries (Barber 2014).

Because of the positive spillover effect subsidies have on related industries, subsidies to a geographically diffuse industry benefit more people than subsidies to a concentrated industry. For example, subsidies to the diffuse construction industry help real estate agents around the country rather than those in just a single city or region, as would be the case if the construction industry was geographically concentrated. In effect, there is a "dispersion bonus" from subsidizing geographically diffuse industries.[13] Parties competing in PR systems profit more from this dispersion bonus than parties in plurality systems. As a result,

---

[13] I am grateful to John Carey for articulating the term "dispersion bonus."

governments in PR systems will spend more on geographically diffuse groups than governments in plurality systems, all else equal.

Even subsidizing a relatively small, yet diffuse, industry could be electorally beneficial in a PR system. If, for example, an industry employs just 2 percent of the population, a party could potentially increase its vote share by 2 percent by subsidizing that industry. Depending on the magnitude of the "dispersion bonus" the electoral gains may be even larger. But even an increase of just 2 percent could be electorally valuable for parties competing in PR systems. In Sweden, for example, a 2 percent increase in a party's vote share could translate into as many as seven additional legislative seats. Given this, I hypothesize that government spending on subsidies will be higher in PR systems than plurality systems when voters with economic interests in subsidies are geography diffuse. In contrast, when the beneficiaries from subsidies are geographically concentrated, subsidy spending will be higher in plurality systems, as compared to proportional systems.

EMPIRICAL EXPECTATIONS

To help generate empirically falsifiable hypotheses, I reduce my argument to a simple two-by-two diagram, illustrated in Table 3.1. I expect to see the most generous subsidies in the cells marked with asterisks. In other words, I hypothesize that governments elected via PR will spend more money on geographically diffuse groups, as compared to governments elected via plurality. This hypothesis is relatively straightforward and follows directly on the logic developed above. Leaders in PR systems have incentives to aid geographically diffuse groups while few politicians do in plurality systems.

Expectations about the political fortunes of concentrated groups are somewhat less clear. On one hand, parties in some PR systems may be indifferent between supporting concentrated and diffuse groups. In PR countries with a single-national electoral district, all votes are equally valuable regardless of their geographic location because they all contribute to a party's share of the national vote. As a result, governments elected via PR from single-national districts may be

Table 3.1 *Illustration of empirical expectations*

	Plurality	Proportional
Concentrated groups	**	
Diffuse groups		**

indifferent between subsidizing concentrated groups and diffuse groups. If so, spending on concentrated groups in single-district PR countries may be similar to that in plurality countries.

Yet, only a handful of PR countries have a single electoral district. Most have multiple electoral districts and legislative seats are often awarded to parties based on their district-level vote shares. This suggests that geography will matter in most PR countries. The question is: how does it matter?

I hypothesize that governments elected via plurality rules spend relatively more on geographically concentrated groups than governments elected via proportionality. Politicians competing in plurality systems have greater incentives to target benefits to groups concentrated in their own districts. Parties in PR systems have few incentives to geographically target benefits in this way. Instead, subsidies for diffuse groups bring greater electoral benefits to parties competing in PR systems, and government parties therefore prioritize funding for diffuse groups over concentrated groups.

Yet, the incentives to target benefits geographically are not entirely absent in PR systems. In some PR systems, namely those with open party lists, individual politicians have incentives to selectively target benefits to groups concentrated their own districts or bailiwicks. Legislators in open-list PR systems seek to divert resources to their own district in order to cultivate their own personal support base. Cultivating a personal vote helps legislators win more individual votes (or preferences) and thereby increases their reelection chances in open-list PR systems. In Chapter 6, I investigate the variation in subsidy spending among PR countries with different types of party lists. I find that although geographic concentration is a political liability in all PR systems, it is relatively less detrimental to interest groups in open-list PR systems. This evidence supports the electoral competition mechanism described above.

Parties competing in PR systems with multiple districts may also seek to target benefits to certain districts. I examine this possibility in Chapter 7 and find evidence that in closed-list PR, governments spend relatively more money on geographically diffuse sectors and simultaneously target subsidies to safe districts. The two strategies are not mutually exclusive. In PR systems, government parties' first-best strategy is to fund subsidies for sectors whose distribution of employees closely matches the geographic distribution of the party's supporters. This strategy is optimal because it is relatively easier to buy the support of voters already predisposed towards the party. Parties therefore seek to fund subsidies whose beneficiaries overlap geographically with the party's supporters or potential supporters (i.e. those people not ideologically opposed to the party). Parties that manage to become the largest party in government in a PR system will tend to have geographically dispersed support. The distribution of employees in geographically diffuse sectors is therefore more likely to match the distribution of the party's supporters. As a result, the party's first best strategy is to target sectors with geographically diffuse employment. Diffuse sectors also provide parties with the widest range of options for geographic targeting. A sector that employs people across the entire country allows parties to selectively target benefits to a particular district via subsidies, as described in Chapter 7.

POSSIBLE ALTERNATIVE MECHANISMS

I argue that two key mechanisms link electoral institutions and economic geography to policy outcomes: the nature of electoral competition and effective vote maximization. However, several other mechanisms may link electoral institutions to policy outcomes. I discuss these possibilities below.

Government Partisanship

Government partisanship may link electoral institutions to policy outcomes. Different electoral institutions may produce different types of governments. Proportional systems, for example, are believed to produce more left-leaning governments than plurality systems (Iversen and Soskice 2006, Rodden 2010). If different electoral systems produce systematically different governments, the mechanism linking electoral institutions to policy outcomes may be government partisanship. Although I take this possibility seriously by controlling for government partisanship in my large-N empirical tests, several factors raise doubts about the plausibility of government partisanship as a key mechanism linking electoral institutions to particularistic economic policies.

First, although partisanship undoubtedly plays a role in economic policy-making, it is unclear how it would affect *particularistic* economic policies, such as subsidies. Previous empirical work finds little evidence of any partisan effects on subsidy spending (Verdier 1995, Rickard 2012c). In fact, no robust correlation exists between government partisanship and subsidy spending in either developed or developing countries (Verdier 1995, Rickard 2012c). Left-leaning governments spend no more (or less) or their total budget on subsidies than right-leaning governments, all else equal.

Second, it is unclear why government partisanship would produce the conditional effects reported in subsequent chapters. The effects of electoral institutions on policy outcomes are conditional on economic geography. This pattern is difficult to credit to government partisanship. Although right-leaning governments may prefer less government spending overall, why they would prefer more (or less) spending on geographically targeted programs, assuming equal costs? The same holds for left-leaning governments.

Third, all governments face demands for subsidies – regardless of their partisan makeup. Many industry groups "hedge their bets" by lobbying governments of all ideological persuasions. In fact, industry groups often pursue friendly relationships with politicians of all parties (Palmer-Rubin 2016). The pursuit of subsidies sometimes engenders unusual alliances that bridge the left/right divide (Thomas 2007). Leaders of industry-based organizations in Mexico, for example, attest to prioritizing a friendly and vigilant relationship with all sitting politicians over declaring firm partisan allegiances (Palmer-Rubin 2016).

These observations suggest that electoral politics may trump partisan considerations.[14] Both left and right parties seek to win office and consequently may provide subsidies under certain conditions. Parties ideologically opposed to subsidies may not have the luxury of taking an ideological stance against them. Parties competing in plurality systems with geographically concentrated industries, for example, would handicap themselves if they refused to provide subsidies. As a result, even right-leaning parties who typically eschew market interventions may find themselves providing subsidies when institutions and geography align. In the United States, for example, one of the most fabled right-wing leaders, President Ronald Reagan, employed industrial policy as a response to what he saw as a loss of US competitiveness in an increasingly globalized world (Foroohar 2017). Reagan agreed to a project, known as "Project Socrates," to study how subsidies helped other nations gain market share in strategic industries with an eye to implementing some of the same strategies in the United States (Foroohar 2017). Similarly, in the United Kingdom, the Conservative Prime Minister Theresa May created a new government ministry responsible for industrial policy. Industrial policy occurs when a government consciously favors some economic activities over others (Rodrik, 2004: 29).

---

[14] Another possible explanation for the null partisanship result may be that changes in government partisanship do not engender changes in economic policies because such policies are sticky. A tariff or subsidy once established becomes "a permanent obligation of government" (Schattschneider, 1935: 92). The longer they have been established the more difficult it becomes to dislodge them (Schattschneider, 1935: 131). Given this, changes in government partisanship may be unlikely to result in changes in subsidies (Thomas 2007).

The minister appointed to head the new Department for Business, Energy & Industrial Strategy said he had been "charged with delivering a comprehensive industrial strategy" (Ruddick 2016). These examples show that even right-leaning political leaders are, in some cases, willing to support market interventions using particularistic economic policies.

Parties across the ideological spectrum respond to the incentives generated by countries' electoral institutions and economic geography. In countries with plurality systems and concentrated economic interests, both left- and right-wing parties have incentives to enact narrowly beneficial economic policies. In PR systems, parties of all stripes have incentives to provide benefits to geographically diffuse interests. This pattern holds even if certain parties have strong historical links to select industries or economic sectors.[15] While parties may support different sectors and industries, all parties in a given country share a willingness to provide subsidies to more (or less) geographically concentrated groups. The incentives for parties to target benefits geographically are generated by the country's electoral institutions. These incentives are constant for all parties in a given country, regardless of their ideology.

Although there are few theoretical reasons to suspect that government partisanship effects subsidy spending, I introduce a measure of government partisanship as a control variable in the cross-national regression models. The inclusion of the partisanship variable does not change the key results regarding electoral institutions and economic geography, which suggests partisanship is not the mechanism linking electoral systems to policy outcomes. As expected, the partisanship variable itself is not consistently robust. In general, left governments do not spend any more (or less) of their budgets on subsidies than right governments. It is possible that left- and right-wing governments assist different industries while spending the same aggregate amount on subsidies. Right-leaning parties may subsidize capital-intensive industries while left-leaning parties may assist labor-intensive industries (Verdier 1995). If different parties subsidize different industries, it is important to account for government partisanship when comparing subsidies between different industries within countries.[16]

---

[15] Agriculture parties, for example, exist in many European democracies. In Norway, the Farmer's Party was established in 1920 to fight for economic policies that favored farmers. In 1959, the party renamed itself the Center Party (*Senterpartiet*) but the party continues to represent agrarian interests (Elder and Gooderham 1978).

[16] However, single-industry studies hold constant industry-specific characteristics such as labor-intensiveness, and as a result, government ideology is not required as a control variable.

Factor Mobility

Another potential mechanism linking economic geography and economic policy is factor mobility. Factor mobility refers to the costs of moving factors of production (i.e. land, labor, and capital) to a new use in the domestic economy. These costs may be, in part, a function of a country's economic geography. Adjustment costs will be higher, for example, when there are fewer alterative employment options in an area. In regions with multiple employers, adjustment costs will be lower.

Adjustment costs have been shown to influence economic policy and more specifically trade policy (e.g. Hiscox 2002, Rickard 2009). However, it is possible that the explanatory power previously attributed to adjustment costs may be due to economic geography. Imagine an electoral district in which the entire working age population is employed in a single sector. Rural districts, for example, are often dominated by agriculture. Nearly the entire working age population may be employed in the agriculture sector or related sectors in a rural district. In such a district, the costs of moving to a new sector are considerable because there are no alternative employers. Moving to a new industry would require a long commute or relocation to a different district. In contrast, workers in a district with many potential employers face lower adjustment costs. In this way, the costs of adjustment are influenced, at least in part, by economic geography. Given this, it is possible that earlier studies that focused exclusively on factor mobility mistakenly attributed causality to mobility rather than geography. I examine this possibility by including a measure of factor mobility in the large-N statistical tests reported in subsequent chapters.

Legislative Dynamics

In my argument, I focus on election dynamics rather than legislative dynamics. I do so for several reasons. First, election dynamics shape legislative dynamics. Evidence in Chapter 5 shows how election dynamics influence a key form of legislative behavior: parliamentary questions. The nature, content, and frequency of questions asked of the government by legislators are shaped by legislators' reelection incentives (Martin 2011a). As the example of parliamentary questions makes clear, to understand legislative behavior, one must understand election dynamics.

Second, legislatures do not always set economic policy. In many countries, cabinet ministers determine key economic policies. In Norway, for example, each ministry prepares its own subsidy budget proposal. The Ministry of Agriculture and Food puts together the budget for subsidies to the agriculture sector. The Ministry for Industry and Trade formulates the budget for subsidies to the manufacturing sector. The proposed budgets are based on input from the various subunits of the ministry and responses from relevant organizations, such as the farmers' lobby groups (*Norges Bondelag* and *Norsk Bonde- og småbrukarlag*) and Innovation Norway (*Innovasjon Norge*), the main bureaucracy responsible for allocating subsidies. The individual ministries' proposals are then put forward to the Ministry of Finance, who prepares the final budget. The government's subsidy budget is presented to the Parliament for approval.

Although Parliament must approve the final budget, individual legislators and opposition parties have little influence over subsidy spending. Parliament "typically does not change the amount of money that has been agreed by the government and interest groups".[17] The government's allocation decisions are normally approved with no amendments or modifications. As a result, ministers enjoy considerable autonomy over the allocation of money to economic sectors that fall within their portfolio. Given this, ministers are uniquely placed to shape particularistic economic policies.

Ministers' policy incentives are largely immune to legislative dynamics. Instead, their decisions are often influenced by electoral dynamics. Many ministers hold office only as long as they or their party win reelection. Even if ministers do not face reelection themselves, they typically work to enact those policies that maximize their party's electoral fortunes in order to keep their party in government. For this reason, I focus on electoral dynamics rather than legislative dynamics.

Legislative dynamics include party discipline, which refers to the ability of a political party to get its legislators to support the policies of the party's leadership. Party discipline is sometimes conflated with the nature of electoral competition.[18] However, the two concepts are distinct. The nature of electoral competition (i.e. candidate versus party centric) is an electoral arena phenomenon while party discipline is a legislative dynamic. Although Cox (1987) proposes the coevolution of these two concepts, party discipline and electoral competition need not be closely related to one another in practice. In fact, only a weak empirical relationship exists between the two (Martin 2011b, 2014). For example, candidate-centered electoral competition occurs in systems with high

---

[17] Siri Lothe, Senior Advisor, Department of Agriculture, Ministry of Agriculture and Food, email communication, July 2, 2015.
[18] For example, party discipline is often estimated using variables that capture the distinction between candidate-centered and party-centered competition (e.g. Ehrlich 2007).

party-discipline, as is the case in Ireland (Martin 2011b, 2014). In short, party discipline and party-centered electoral competition are two distinct concepts that do not always go together.

Just as party discipline does not perfectly correlate with the nature of electoral competition, neither does it vary systematically with electoral institutions. Within plurality systems, there are both high and low discipline systems (McGillivray 1997, 2004). National parties in Canada, for example, are stronger than parties in the United States – even though both countries use plurality electoral rules. In systems like the US, where party discipline is low, legislators can represent the interests of their constituents even if they go against the interests of their party (Schattschneider 1935). Variation in party discipline also exists among PR systems (Martin 2011b, 2014). Some PR systems have relatively low levels of party discipline while others have much higher levels of party discipline. The variation suggests that electoral rules by themselves fail to fully explain party discipline.

In short, the origins of party discipline are unclear. While electoral systems may have some effect on party discipline, many other factors matter as well, including a party's organizational structure and the allocation of power inside legislatures. These factors explain why party discipline is largely uncorrelated with electoral rules. Given this, party discipline cannot be the mechanism linking electoral systems to policy outcomes. In fact, previous studies show that party discipline has no robust effect on trade policy after controlling for electoral rules (Hatfield and Hauk 2014). This evidence points to the relative importance of electoral dynamics.

Electoral competitiveness

Given my focus on electoral dynamics, the competitiveness of elections is potentially important. Electoral competitiveness typically refers to the closeness of an election in a given district. The competitiveness of elections often varies between districts in countries. In some districts, incumbents face tough electoral competition while in others incumbents face no serious challenger.

While undoubtedly important, district-level electoral competitiveness is an unlikely explanation for the variation in economic policies across countries. First, district-level competitiveness cannot be meaningfully aggregated to the country level because the competitiveness of elections varies between districts within countries. In practice, of course, one could calculate the "average" level of district-level competitiveness in a country; however, it is unclear what, if anything, this aggregate value would mean – particularly in a cross-national context. Second, proportional systems do not have marginal districts in the same sense as plurality systems; in PR systems, districts are multimember and seats are distributed based on the proportion of votes within each district (McGillivray, 2004: 19). Because competitiveness means different things in different electoral systems, it is hard to imagine how competitiveness could explain the varied policy outcomes between countries with different electoral institutions. Third, average competitiveness levels vary within electoral systems (Kayser and Lindstädt 2015). Elections are no more (or less) competitive, on average, in plurality systems than in proportional systems (Kayser and Lindstädt 2015). Given this, competitiveness is unlikely to be the causal mechanism linking electoral institutions to national policy outcomes.

Although district-level electoral competitiveness is an unlikely explanation for the variation in subsidies between countries, it may help to explain the variation in subsidies within countries. District-level electoral competitiveness may explain why some electoral districts receive more generous subsidies than others, controlling for economic geography.

A large literature examines the role of district-level electoral competitiveness for policy targeting in plurality countries (e.g. Cox and McCubbins 1986, Dixit and Londregan 1996). Although scholars disagree about precisely which districts (i.e. safe or swing districts) receive the most generous fiscal benefits or "pork," scholars agree that politicians have electoral incentives to target benefits geographically in plurality systems. These arguments coincide with my own; I argue that plurality electoral rules incentivize the targeting of economic benefits to geographically concentrated groups. Whether parties target safe or swing districts is immaterial for my theoretical argument. What matters is that politicians and parties target benefits to geographically defined electoral districts because of the incentives generated by plurality rules. I remain agnostic as to whether leaders in plurality systems target safe or swing districts. I do not empirically examine the variation in economic rents between districts within plurality countries because many previous studies have already carried out this task.[19]

In Chapter 7, I examine how electoral competitiveness influences the variation in subsidies between electoral districts in a PR country with de facto closed party lists, controlling for economic geography. I find evidence that the largest party in government targets subsidies disproportionality to districts where it received a greater share of the

[19] For a comprehensive review of this literature see Golden and Min (2013).

vote in the previous election. In other words, incumbent parties in closed-list PR systems appear to target economic benefits to "safe" districts. This result suggests that electoral competitiveness can help to explain the variation in economic rents between districts within PR countries.

Of course, electoral competitiveness itself may be a function of economic geography (Kayser and Lindstädt 2015, Rodden unpublished manuscript).[20] Certain electoral districts are "safe" for some parties precisely because of their economic characteristics. In Britain, for example, the Labour Party historically dominated elections in districts where coal mining was the main economic activity (Johnston et al. 2006). Labour won these seats with large majorities, thanks to the high voter turnout stimulated by miners' trade unions (Johnston et al. 2006). As the number of people employed in the mining industry declined, turnout fell in a number districts that the Labour Party had previously considered "safe" (Johnston et al. 2006). As a result, the magnitude of Labour's margins declined and the party began to face more competitive races in these districts. The experience of Britain's Labour Party shows how economic geography matters for electoral competitiveness. My argument focuses on the primary causal factor: economic geography.

In most of the book, I focus on the variation in economic policies between democracies with different electoral systems. I argue that spending on subsidies for geographically concentrated groups will be higher, on average, in plurality systems than proportional systems. District-level electoral competitiveness may influence which districts receive more economic rents but it will not alter the fact that politicians in plurality systems have greater incentives to target benefits geographically, as compared to politicians in PR systems. In Mexico, for example, where most legislators are elected via plurality in single-member districts, the identity of subsidy recipients changes depending on the competitiveness of state elections (Palmer-Rubin 2016). But regardless of the competitiveness of state-level elections, political parties target selective benefits to geographically concentrated groups (Palmer-Rubin 2016). In other words, electoral competitiveness does not influence aggregate policy targeting or the overall generosity of government subsidies. Instead, it affects the identity of subsidy recipients within a country. Given this, electoral competitiveness may more usefully explain policy

---

[20] Kayser and Lindstädt (2015) report that the effect of electoral rules on real prices is mediated by a variable they label "loss probability." The construction of their "loss probability" variable depends critically on geography (and highly restrictive assumptions for PR countries). Their results can therefore be interpreted as evidence that geography mediates the effect of electoral rules on economic outcomes. In other words, the results of Kayser and Lindstädt (2015) are consistent with my argument.

variation *within* countries rather than across countries, as I explore in Chapter 7.

CAUSAL COMPLEXITY

In this book, I take political institutions and economic geography as given (i.e. exogenous). In other words, I do not speculate as to where electoral institutions come from or what causes certain patterns of economic geography. Because I focus on policy-making in the short to medium term, this choice makes sense. In the long run, however, both institutions and economic geography may evolve over time and these changes might be systematically related to one another.

Economic geography may influence the design of electoral institutions. Institutions are, after all, "humanly devised constraints" that may be intentionally constructed to obtain certain purposes (North 1990). But once chosen, electoral institutions are "sticky" and tend to change only rarely over time. In the short to medium term, the geographic distribution of sector-specific employment is unlikely to influence these decidedly durable institutions. However, in the long run, the spatial distribution of economic activity may influence a country's choice of electoral system (Cusack, Iversen, and Soskice 2007). If the geographic distribution of economic activity is influenced by subsidies and the geography of economic activity influences electoral system design, then subsidies may shape countries' electoral systems in the very long run. If true, this would raise questions about the logic of my argument and the validity of my empirical tests. However, I find no evidence that subsidies or economic geography influence countries' decisions about electoral systems.

No significant difference exists between the geographic concentration of manufacturing employment in PR systems and plurality systems.[21] Furthermore, changes in the geographic concentration of economic interests have not been accompanied by corresponding changes in electoral institutions. In fact, the handful of recent changes in electoral institutions have had little, if anything, to do with economic geography. In Italy, for example, the move from PR to plurality that occurred in 1993 was driven by a desire to "clean up" Italian politics and increase the durability of governments (Katz 1996). Similarly, New Zealand's electoral system reforms were motivated by myriad issues – none of which related to economic geography. These observations minimize

---

[21] The sample mean value of *Concentration* is equal to 0.035 in PR countries and 0.031 in plurality countries. The difference (−0.004) is not statistically significant, as demonstrated by a two-sample t-test with equal variances.

concerns about the influence of economic geography on electoral institutions.

Economic actors might strategically locate in response to electoral institutions and if so economic geography could be a function of a country's electoral system. If actors anticipate the benefits of geographic concentration in plurality systems, for example, they may choose to locate in districts with other similar actors to maximize their chances of winning economic support from the government. But the observed geographic concentration of manufacturing is no higher in plurality systems than in PR systems, on average, which suggests that location decisions are not driven by electoral institutions. In fact, no evidence exists to suggest that firms strategically locate in response to a country's electoral rules.

Subsidies generally do little to influence firms' location decisions (Midelfart-Knarvik, Helene, and Overman 2002). Patterns of geographic concentration are relatively immune to government spending programs. Large-N statistical tests show that subsidies have negligible effects on firms' location decisions (Devereux et al. 2007). Furthermore, civil servants report that they have never seen a firm reallocate to try to win more subsidies from the government. Bureaucrats in Norway, for example, acknowledged that it was more difficult for a firm in the capital city of Oslo to win subsidies because of the higher concentration of firms in that district. Norway uses proportional electoral rules so this statement is fully consistent with my argument: geographic concentration confers a political handicap on groups in PR systems. Yet, no firms moved out of Oslo to increase their chances of winning a government subsidy.[22] Taken together, these observations minimize concerns about causal complexity. In the short to medium term, electoral institutions and economic geography appear to be exogenous to each other and to government-funded subsidies. However, to address any lingering concerns about causal complexity, I employ instrumental variables in my quantitative empirical tests and carefully evaluate the qualitative evidence for evidence of reverse causality.

CONCLUSION

Economic geography mediates the effects of electoral institutions on economic policy. While electoral institutions define the optimal (re)election strategy for politicians and parties, economic geography

---

[22] In person interview with Innovation Norway staff members Pål Aslak Hungnes and Per Melchior Koch in Oslo, Norway on June 19, 2014.

Conclusion

determines how best to achieve this strategy. Plurality electoral rules incentivize policies that target benefits narrowly to precise geographic locations. Subsidies can be used to achieve this goal – but only when the beneficiaries are geographically concentrated. When they are geographically diffuse, subsidies are an inefficient electoral tool in plurality systems. Therefore, I hypothesize that geographically concentrated groups will win more generous subsidies in plurality systems as compared to proportional systems, holding all else equal.

In PR systems, parties seek to maximize the number of seats they control in the legislature. Subsidies for geographically diffuse groups are an effective means to achieve this goal. As a result, I hypothesize that geographically diffuse groups will win more generous subsidies in proportional systems as compared to plurality systems, holding all else constant. In the following chapter, I test these hypotheses empirically using cross-nationally comparable data on government spending on industrial subsidies.

4

Explaining Government Spending on Industrial Subsidies

In plurality systems, geographically concentrated interest groups enjoy a political advantage because of the (re)election incentives politicians face. Politicians win office by securing a plurality of votes in their electoral district. To achieve this goal, politicians target benefits to groups concentrated in their district. In contrast, politicians in proportional systems have fewer incentives and little opportunity to selectively target benefits to their own district. Politicians in PR (proportional representation) systems work instead to garner votes for their political party. Parties work to assist geographically diffuse groups to maximize the number of seats they control in the legislature. Given these incentives, I hypothesize that geographically diffuse interest groups will enjoy greater political influence in proportional electoral systems, as compared to plurality systems.

Although the logic of my argument is general and can be applied to a variety of policy areas, I focus here on economic policy, and specifically on government-funded subsidies. A subsidy is a financial contribution by a government or agent of a government that confers a benefit on its recipients. Although governments subsidize many types of different groups, I focus here on government subsidies to businesses and specifically to producers. I investigate how much money governments spend on producer subsidies. Government spending on subsidies provides an expedient test of my argument. Nothing speaks louder about a government's priorities than how it spends its money. The more money governments spend on producer subsidies, the less money is available for other programs, such as education and health care. As an economist in the Indian government said, "every subsidy I am giving is money that the government could have spent elsewhere. Every subsidy means a primary healthcare centre I cannot build."[1]

---

[1] Economist Bibek Debroy from the National Institution for Transforming India, a Government of India policy think-tank. Quoted in the Economic Times, India

Recipients of producer subsidies vary meaningfully across space; some recipients are geographically dispersed across an entire country while others are concentrated in small geographic areas. Given this, producer subsidies are almost infinitely malleable (Verdier 1995). They can be more or less narrowly targeted depending on the geographic distribution of potential beneficiaries and the eligibility criteria fashioned by the government. Governments can make subsidies automatic in nature with few, if any, eligibility criteria. For example, a government could subsidize the cost of energy for all businesses, regardless of their geographic location or sector. Such subsidies would be broadly beneficial. Alternatively, governments could impose restrictive eligibility criteria that effectively limit subsidies to select producers. Such targeted, selective programs are sometimes referred to as "pork barrel" spending Persson and Tabellini (2003: 14).

In this chapter, I focus on subsidies targeted by the government to the manufacturing sector. Manufacturing subsidies benefit people employed in the manufacturing sector. To ascertain the geographic distribution of these beneficiaries, I measure the geographic concentration of manufacturing sector employment. The geography of manufacturing employment varies between countries and this cross-national variation allows me to empirically test my argument about the interactive effect of geography and institutions on particularistic economic policy.

MEASURING PARTICULARISTIC ECONOMIC POLICY

To test my argument, I examine government spending on subsidies in developed countries. Specifically, I focus on the amount of money governments spend on subsidies for the manufacturing sector. Before reporting my empirical findings, I begin by justifying each part of my empirical strategy. First, I describe what subsidies are and why they are worthy of investigation. Second, I explain my focus on sector-specific subsidies – that is subsidies provided to a specific sector of a country's economy. Third, I explain why I focus on subsidies to manufacturing. Finally, I discuss the advantages of subsidies, as compared to other types of government spending, as well as potential drawbacks.

Times, May 7, 2017. "Scrap all production subsidies" http://economictimes.indiatimes.com/news/economy/policy/scrap-all-production-subsidies-niti-aayogs-bibek-debroy/articleshow/58560312.cms.

What Are Subsidies?

Disagreement abounds about the precise definition of subsidies. "The word subsidy derives from the Latin word *subsidium*, which means 'help, support, or assistance' and in medieval times referred to a payment made to the king" (Fryde 1983, Steenblik 2017).[2] Today, subsidies cover a wide range of government policies. Subsidies include: cash grants, tax breaks, loans at below-market interest rates, loan guarantees, capital injections, guaranteed excessive rates of profit, below-cost or free inputs including land and power, and purchasing goods from firms at inflated prices.[3] The breadth of possible policy options generates confusion about exactly what constitutes a subsidy. The World Trade Organization provides one of the only internationally agreed upon definitions, which delineates a subsidy as a financial contribution by a government that confers a benefit on its recipients (WTO 2006). I adopt this definition.

Just as many definitions of subsidies exist, there is also a striking array of synonyms for subsidies. Frequently used terms include industry assistance, industrial policy, industrial strategy, corporate welfare, and government support programs. The European Union uses the term "state aid" to describe subsidies provided to producers by member state governments. Another increasingly common term is "economic incentive" (e.g. Jensen 2017), which typically refers to subsidies that take the form of a tax incentive. Tax incentives reduce companies' tax burdens thereby allowing businesses to keep a larger share of their revenue. Quantifying tax incentives is notoriously difficult and few, if any, cross-nationally comparable measures of tax incentives exist.

Like tax incentives, many other types of subsidies are also difficult to measure. Such difficult-to-measure subsidies include those provided through government procurement (e.g. Rickard and Kono 2014, Kono and Rickard 2014), policies that raise prices artificially (i.e. market price support), credit subsidies and government guarantees – just to name a few. An illustrative example of a difficult-to-measure subsidy comes from Norway, where the national government provides e-certificates to renewable energy producers.[4] Producers earn money from selling the e-certificates to electricity suppliers who then pass on the costs to

---

[2] www.iisd.org/gsi/getting-know-subsidies. For example, King Edward I of England provided large subsidies to potential enemies of France (Fryde, 1983: 1170). The most generous subsidies were paid by Edward I to the German king Adolf of Nassau (Fryde, 1983: 1170).
[3] This list is not exhaustive, but includes the most frequent types of support.
[4] www.regjeringen.no/en/topics/energy/renewable-energy/electricity-certificates/id517462/.

consumers. These e-certificates are valuable; they were estimated to be worth 0.7 billion Norwegian krone (NOK) in 2013 and are expected to increase in value to 2.8 billion NOK by 2020.[5] Via these e-certificates, the government confers a valuable benefit on producers without incurring a financial liability. As a result, the subsidies provided via the e-certificates do not show up on the expenditure side of the government's budget. Such subsidies are extraordinarily difficult to identify and measure – especially in a systematic and cross-nationally comparable fashion.[6]

To sidestep these difficulties, I measure subsidies that generate financial commitments for governments and are therefore visible in government budgets. Subsidy schemes on the expenditure side of the fiscal budget include direct payments, grants, loans, or guarantees granted on preferential terms. A grant refers to a time-limited payment, either in connection with a specific investment, or to enable an individual, company, or organization to cover some or all of its general costs, or costs of undertaking a specific activity, such as research (Steenblik 2017)[7]. Other direct payments to producers may be linked to the volume of production or sales. By measuring subsidies that show up on expenditure side of the fiscal budget, I construct a cross-nationally comparable measure of governments' subsidy spending. This measure, however, represents a lower bound for total government support for producers.

Why Study Subsidies?

Subsidies are worthy of investigation because they have profound and long-lasting effects on the economy, the distribution of income in society, and the environment (Steenblik 2017).[8] Subsidies also provide valuable information about questions that lie at the heart of democracy: how responsive are politicians to narrow interest groups? When do governments privilege the interests of a few over the good of many? Government spending on subsidies and the variation in subsidy spending between democracies can help to provide answers to these important questions.

Democratically elected governments spend a sizeable amount of money on subsidies for business. In the European Union, subsidies to the

---

[5] Lone Semmingsen, Deputy Director General, Ministry of Finance, Norway, Written communication, January 20, 2015.
[6] However, the correlation between expenditure subsidies and total subsidies in Norway equals 0.6.
[7] www.iisd.org/gsi/getting-know-subsidies.
[8] www.iisd.org/gsi/getting-know-subsidies.

manufacturing sector accounted for 2 percent of value added in the sector in 2003 or approximately €1,000 per person employed in manufacturing (Sharp 2003).[9] In 2003, twenty-one developed countries spent nearly $250 billion on subsidies. Subsidies likely account for an even greater share of government expenditures in developing countries (IMF 2001, Fan and Rao 2003). In India, subsidies amounted to around 14 percent of GDP in 2016.[10] Total global spending on subsidies was more than $300 billion in 2003 and since then, subsidy spending by governments around the world has increased (Thomas 2007). Given the prominence of subsidies in governments' budgets and their significant effects, it is important to understand the politics behind these programs.

Governments spend more and more money on subsidies in response to the growing number of international agreements that limit tariffs. To assist domestic producers that face growing competition from foreign imports as a result of lower tariffs, governments often provide subsidies (Ford and Suyker 1990, OECD 1998, Rickard 2012b). For example, the Japanese government promised to increase subsidies to pig farmers in order to offset the reduction in pork tariffs agreed as part of the Trans-Pacific Partnership (TPP) multilateral free trade agreement.[11] As this example illustrates, governments may provide subsidies as a substitute for tariffs (Rickard 2012b). Subsidies are, in fact, one of the few tools policy-makers have remaining at their disposal to assist domestic industries. World Trade Organization (WTO) commitments restrict the use of tariffs and capital markets restrict exchange rate manipulation (Blomström, Kokko, and Mucchielli 2003, Thomas 2007). Given these international restrictions, subsidies may become the universal mode of state intervention in industry (Verdier 1995).[12]

Subsidies reveal valuable information about governments' spending priorities. Subsidies typically benefit a minority group (i.e. an industry or sector) at the expense of a majority of citizens (taxpayers). In other words, subsidies tend to provide concentrated benefits with diffuse costs. Subsidies to the US sugar industry, for example, uphold a domestic sugar

---

[9] Down from 3 percent in 1995 and 4 percent in 1990.
[10] This estimate includes budgetary and off-budgetary subsidies and those relating to consumption and production and comes from the National Institution for Transforming India, a Government of India policy think-tank. Times, May 7, 2017, "Scrap all production subsidies" http://economictimes.indiatimes.com/news/economy/policy/scrap-all-production-subsidies-niti-aayogs-bibek-debroy/articleshow/58560312.cms.
[11] The Japan News, May 21, 2014, S Edition, Business Section, p. 8 (accessed via Lexis Nexis).
[12] Some international restraints exist on subsides. However, theses "are either voluntary or do not bind in a significant way" (Rodrik, 2004: 32).

price two to three times higher than the world's market price (Mortensen and Pissarides 2001). As a result, sugar cane farmers in the United States receive, on average, an extra $369 million a year above the internationally determined value for the commodity (Beghin et al. 2003, Frieden et al., 2010: 234).[13] These benefits come at a cost to American taxpayers and consumers who pay $2.3 billion a year more for sugar due to government subsidies and other economic policies designed to support the industry (Beghin et al. 2003).

Because subsidies, like those to the US sugar industry, benefit select groups at the expense of many, government spending on subsidies provides valuable information about how leaders weigh narrow demands against broader societal interests. This is especially true when subsidy spending is reported as a percentage of total government expenditures, as it is here.[14] The ratio of subsidy spending to total government spending indicates the relative importance of subsidies among governments' myriad spending priorities. This ratio provides a novel, direct measure of how politicians weigh narrow demands against broader societal interests.

Estimating governments' responsiveness to narrow or special interests is notoriously difficult. Imperfect proxies have been used in earlier studies. National price levels, for example, have been used to estimate the responsiveness of politicians to narrow producer interests (Rogowski and Kayser 2002). Similarly, industry stock prices have been employed to measure industries' political influence (McGillivray 2004). These indirect measures, while creative, capture many factors that have nothing to do with governments' responsiveness to narrow interests, such as transportation costs, market size, and consumer demand (Rodriguez and Rodrik 2000).

To estimate how elected leaders weigh narrow demands against broader societal interests, some scholars investigate government spending patterns. However, the spending measures examined to date are far from ideal. Spending on social security payments and other transfers to families, plus subsidies to firms, has been used as a proxy

---

[13] Calculations are for 1998 converted into 2006 US dollars.
[14] Excluding interest payments. These spending data come from the International Monetary Fund's *Government Financial Statistics*. These data include all fiscal outlays targeted to the manufacturing sector. For example, all subsidies, grants, and subsidized loans provided to the manufacturing sector to support manufacturing enterprises and/or development, expansion, or improvement of manufacturing are included. Although conventional government accounts are generally not suitable for comparisons among countries and over time because they reflect the organizational structures of governments, these data, uniquely allow meaningful cross-national comparisons over time. For additional information, see IMF (2001).

narrowly targeted transfers (Milesi-Ferretti, Perotti, and Rostagno 2002). Broadly targeted transfers have been estimated using government spending on education, transportation, and order and safety (Persson and Tabellini 1999). Yet, both measures conflate programs that may be more or less narrowly targeted depending on the precise eligibility criteria outlined by the government. In contrast, subsidies provide a more accurate measure of governments' spending priorities. Using subsidies, it is possible to know exactly how targeted a given government program is by identifying the subsidy's eligibility criteria and the geographic distribution of eligible recipients. In sum, subsidies present a novel way to measure governments' responsiveness to narrow interest groups, which may or may not be geographically concentrated.

Why Focus on Sector-Specific Subsidies?

In this chapter, I focus on sector-specific subsidies – that is, subsidies targeted to a specific sector of a country's economy. Governments often specify their subsidy priorities in terms of sectors (Rodrik, 2004: 23; Chapter 7). Sector-specific subsidies are often voted on by legislatures and consequently provide a useful test of my theory. What's more, sector-specific subsidies are comparable between countries because of internationally accepted accounting rules. International accounting rules require governments to report total budget support to certain sectors of the economy, including manufacturing (IMF 2001). As a result, it is possible to measure how much governments spend on sector-specific subsidies in a cross-nationally comparable manner. In contrast, it is virtually impossible to measure government support for individual firms or industries in a cross-nationally comparable fashion. Governments have no international obligations to report how much money they spend on a given firm or industry. And individual industries are often aided by multiple budget items. The Australian wine industry, for example, was assisted by at least four different items in the 2003–2004 national budget including drought assistance, preferential excise rate, tax deductions for grape vines, and direct fiscal transfers to the Grape and Wine R&D Corporation.[15] These support programs are scattered throughout the budget and are managed by different government ministries.

Given the multitude of budget items that could potentially support a single industry, it is difficult, and arguably impossible, to calculate the total amount of financial support provided to an industry in a given country. As the UK government admits: "[t]here is no definitive source

---

[15] Author's calculations from government budget documents.

of data about spending on subsidies to businesses in the UK."[16] As a result, aggregating total governmental budget support for individual industries in a cross-nationally comparable manner is virtually impossible. This empirical challenge likely explains the dearth of large-N, cross-national studies of government aid to individual industries.[17] In contrast, international accounting rules make it possible to compare spending on sector-specific subsidy programs between countries. For this reason, I focus on government spending on sector-specific subsidies in this chapter.

Governments in many countries have direct control over the amount of money spent on sector-specific subsidies. Legislatures often vote on how much money to allocate to particular sectors of the economy in the annual budget. In Norway, for example, the government specifies how much money will be allocated to the agriculture sector, the manufacturing sector, etc., in the annual budget, which is subsequently voted on by parliament.[18] Similarly, in Mexico, the initial allocation of funding for subsidies to the agriculture sectors is made by Congress.[19] Because national legislatures often decided how much money to spend on sector-specific subsidy programs, the political dynamics I theorize about likely impact sector-specific subsidy spending. In contrast, most governments

---

[16] www.theguardian.com/politics/2015/jul/07/corporate-welfare-a-93bn-handshake.
[17] Of course, some efforts have been made to overcome this data limitation. For select agricultural products, the OECD calculates producer support estimates and Anderson and Swinnen (2008) calculate "nominal rates of assistance." While these data are commendable, their usefulness for my purposes is limited. First, both measures are limited to agriculture products only, while my focus here is on manufacturing. Second, not all products are covered for all countries. Nominal rates of assistance were estimated for less than a dozen products per country on average (Anderson, 2009: 7). Wine, for example, is covered in some, but not all European countries. Wine is not covered in countries such as Australia, New Zealand, Chile, or South Africa. The lack of coverage does not reveal an absence of government assistance. The Australian wine industry, for example, is heavily subsidized via a multitude of government programs. Yet, wine is not included in either the OECD dataset or in Anderson and Swinnen (2008). Given the limited coverage, it is difficult to use these data to make cross-national comparisons of support to an individual industry or producer group. Finally, the measures of nominal rates of government assistance include non-product-specific forms of assistance or taxation in addition to industry or product specific aid. By combining both forms of assistance, these measures cannot be used to draw inferences about government support targeted to a single producer group.
[18] Once the sector-specific budget allocations have been set by the government, bureaucrats disperse the money. Bureaucrats are, however, constrained by elected leaders, as described in Chapter 7.
[19] The State Councils for Sustainable Rural Development (CEDRS) are formally vested with the authority to decide how to allocate agriculture-sector program resources (Palmer-Rubin 2016).

only have indirect control over the amount spent on subsidies for individual firms or industries. For these reasons, I focus here on sector-specific subsidies and how they vary between countries.

Why Manufacturing?

In this chapter, I focus on subsidies provided by governments to the manufacturing sector. I focus on the manufacturing sector for several reasons. First, data on manufacturing-sector subsidies are widely available and comparable between countries. Second, most research to date focuses exclusively on agriculture subsidies, and as a result the substantial variation in *manufacturing* subsides has gone largely unexamined. Third, governments in many developed countries face increasing demands to assist the manufacturing sector as employment in the sector declines. Government support for manufacturing was a key issue in the 2016 US presidential election campaign. Donald Trump's promise to bring "manufacturing back to America"[20] helped him win votes in historically Democratic areas and ultimately the election. As more and more governments come under pressure to aid manufacturing, my theory can help to explain why governments in some countries choose to spend more money on manufacturing aid than others.

Benefits and Weakness of Subsidy Measure

Examining subsidies brings many benefits. Subsidies illustrate governments' spending priorities and have long-lasting effects on a country's economy and the distribution of income within society. Yet, no empirical measure, including subsidy spending, is without weaknesses. One potential drawback of studying subsidies is the increasing number of international restrictions on subsidies. International agreements, including the EU and WTO, increasingly restrict governments' use of subsidies. International restrictions on subsidies may make it difficult to observe the domestic politics behind such programs. Although governments often have electoral incentives to provide subsidies, they may stop short of doing so because of international restrictions. This type of behavior would make it difficult to find evidence in support of my argument. In other words, international rules regulating subsidies may bias against finding any evidence of the theorized interactive effects of domestic institutions and economic geography.

---

[20] www.whitehouse.gov/the-press-office/2017/02/23/remarks-president-trump-meeting-manufacturing-ceos.

In practice, however, many governments continue to subsidize domestic producers – even as members of the EU and WTO (OFT 2009). One reason for the continued use of subsidies is that most international restraints on subsides are "are either voluntary or do not bind in a significant way" (Rodrik, 2004: 32). As Rodrik points out, "[i]t is easy to exaggerate the significance of the [international] restrictions [on subsidies]. There remains much scope for coherent industrial policy" (Rodrik, 2004: 32). Most international agreements provide exemptions that allow for certain types of subsidies, including subsidies for research and development, small and medium-sized enterprises, innovation, training, regional development, employment, environmental protection, as risk capital, and for promoting entrepreneurship (OFT 2007). EU state aid rules also allow governments to provide subsidies to failing firms in the form of rescue and/or restructuring aid (European Commission 2009). Government use these exemptions to subsidize domestic producers. Their willingness to do so, I argue, depends on a country's electoral institutions and economic geography.

MEASURING ECONOMIC GEOGRAPHY

Given the strikingly uneven dispersion of economic activity, it is surprising how few empirical measures of economic geography exist. But measuring geographic patterns of economic activity is difficult. Doing so requires vast amounts of highly disaggregate data. Information about producers' geographic location is needed and such data are often unavailable or confidential – even in an era of ever increasing data transparency.

An illustrative example of the challenges involved in measuring economic geography comes from Austria – where measuring the geographic distribution of even a single industry proved difficult. I set out to measure the geographic distribution of Austrian wine production by collecting data on the geographic location of wine producers. Neither the government nor the industry itself had accurate data on the location of wine producers. A comparison of the wine industry organization's data for Vienna (*Wien*) in 2014 with data from the municipal authority revealed myriad discrepancies in the location of production sites.[21] As this example illustrates, it is difficult to measure economic geography – even for a single industry in one country.

Going forward, improvements in data transparency may make it easier to accurately measure economic geography. In Austria, for example, the wine marketing board wrote in 2014 that it hoped to have a new database

[21] Austrian Wine Statistics Report 2014.

"based on a geographical information system which will allow an annual snapshot to be taken of the [geographic] distribution by grape variety in Austria's wine regions ... over the next few years."[22] Although this goal has yet to be fully realized, economic geography will likely become easier to measure as more accurate geographic data are compiled and shared.

Improved data-sharing among European countries allowed me to estimate the geographic distribution of manufacturing employment using entropy indices, which I describe in detail below. First, I begin with a brief discussion of the handful of existing empirical measures of economic geography.

Early studies used Gini indexes to measure the geographic concentration of an industry's regional and national employment shares (e.g. Krugman 1991). A Gini coefficient measures the inequality among values of a frequency distribution (for example, levels of employment in regions). A Gini coefficient of zero expresses perfect equality, where all values are the same while a coefficient of one expresses maximal inequality among values. Following Krugman (1991), Gini indices became the standard measure of geographic specialization patterns. The Gini index has strong intuitive appeal. These indices are straightforward to compute and relatively manageable in their data requirements. However, Gini indexes are limited in their usefulness because they cannot be decomposed into within-country and between-country components. The Gini index is only decomposable if the range of the values taken by the variable of interest does not overlap across subgroups of individual observations (Cowell 1980).[23] Yet, regions in different countries may have similar degrees of specialization in a particular sector. As a result, it is often not possible to accurately construct within-country and between-country components of geographic concentration using Gini indexes. This challenge may explain why virtually all existing research on the political consequences of geographic concentrated using Gini indexes is limited to single-country studies. To compare the effects of geographic concentration between different countries, decomposed values are needed.

Improvements on the Gini indexes emerged from the growing literature on economic geography. One such improvement was the "dartboard approach," developed by Ellison and Glaeser (1997). This approach controls for variation in the size distribution of plants in an industry and the size of geographic areas. However, this approach cannot distinguish between industries with concentration in parts of a country far apart from

---

[22] Austrian Wine Statistics Report 2014.
[23] Also see Cowell (1995) for a discussion of problems related to the Gini.

one another or in contiguous areas. For example, this approach could not distinguish an industry with employment in northern Wales and southern England from one with employment in two contiguous counties near London, even though the latter is clearly more concentrated topographically – a problem often referred to as the "checkerboard problem."

Some scholars have attempted to address the "checkerboard problem." For example, Busch and Reinhardt (2000) calculate geographic concentration by measuring the distance between each employee and the national "centroid" or midpoint for a given industry (p. 708). However, the center of a given industry is determined in an ad hoc fashion, which raises questions about the construction of this variable. Furthermore, calculating this variable requires a vast amount of data. As a result, it exists for only two countries both of which have plurality electoral rules: the United States and the United Kingdom.

Most recently, entropy indices have been used to measure geographic concentration.[24] These indices have several distinct advantages over earlier measures. One key advantage lies in their suitability for decomposition analysis. Using entropy indices, it is possible to compute within-country sector concentration measures in a conceptually rigorous fashion. Using entropy indices, I calculate the geographic concentration of manufacturing employment. More precisely, I calculate the degree to which employment in a country's manufacturing sector is geographically concentrated compared to a "no-concentration" benchmark. Two possible no-concentration benchmarks can be used. The first no-concentration benchmark is calculated by dividing the sector's total employment by the number of subnational units, or regions, in a given country. The resulting value reports how many people would be employed in each subnational unit if sector employment was equally distributed within the country. This value serves as a baseline no-concentration benchmark. For example, in 2014 there were 83,779 employees in the Norwegian petroleum and petroleum-related industries (Statistics Norway 2015). If employment in this sector was evenly dispersed across the country's nineteen electoral districts, each district would have 4,409 petroleum employees. This number is the no-concentration benchmark for topographical concentration. The no-concentration benchmark obtains where an activity is spread evenly over physical space. Any departure from such an even spatial spread will register as geographic concentration, irrespective of the spatial distribution of

---

[24] Prior to Brülhart and Traeger (2005), the use of entropy indices in studies of economic geography was relatively rare. However, entropy indices were common in the income distribution literature (see e.g. Cowell, 2000).

endowments or of other economic sectors. I refer to this concept of geographic concentration as "topographic" concentration.

However, economic activity is not equally distributed topographically. In Norway, for example, some regions are sparsely populated and have little employment or economic activity of any kind. To account for this, I calculate a second measure of geographic concentration where I weight sector employment by total employment in each region. In other words, I condition physical space by the distribution of aggregate employment. If employment in the petroleum industry was perfectly distributed relatively to total employment, it would account for 5 percent of employment in each electoral district in 2014 (Statistics Norway 2015). This value represents the no-concentration benchmark. In reality, however, the petroleum industry it is concentrated geographically. In the western region of Rogaland, for example, the petroleum industry employs nearly 40 percent of the population. The deviation from the no-concentration benchmark indicates a high level of "relative" concentration. Relative concentration measures the degree to which sector employment is concentrated *relative* to the geographic distribution of aggregate employment. This concept of concentration is often (implicitly) invoked when theorizing about the political effects of economic geography. I use relative concentration measures throughout this book given its political relevance.[25]

While entropy indices have appealing characteristics, their computation requires disaggregate sector-specific employment data for subnational geographic units. These data are only rarely available in developed countries (and virtually nonexistent in developing countries). Given the large amounts of disaggregated data needed to construct entropy indices, my sample is necessarily limited. It includes only those countries that report disaggregated employment data for subnational regions in a cross-nationally comparable format. Fortunately, the European Union requires member-states to report data on employment by sector and subnational regions in a cross-nationally comparable format to Eurostat. These reporting requirements make possible the construction of entropy indices for fourteen European countries over two decades. Although this limited sample raises potential questions about the external validity of the results, it allows for direct comparisons with previous studies of electoral institutions that use similar samples (e.g. Bawn and Rosenbluth 2006,

---

[25] Although relative concentration captures a theoretically different concept than topographic concentration, the two are often highly correlated. In Norway, for example, the two measures of concentration are correlated at 0.9 over the period from 1999 to 2014.

Persson, Rolland, and Tabellini 2007). As data transparency improves, it may be possible to construct entropy indices for additional countries in the future.

To calculate country-specific sector concentration measures, I use the following equation:

$$\text{GE}(1)_s = \frac{1}{N} \sum_{i=1}^{N} \frac{y_{si}}{\bar{y}_s} \log \frac{y_{si}}{\bar{y}_s}$$

where N is the number of NUTS-2 units in a given country. NUTS units refer to subnational geographic areas defined by the European Union's Nomenclature of Territorial Units for Statistics (NUTS). This classification scheme was designed for the purpose of the collection, development, and harmonization of statistics within the union, which makes it ideal for constructing cross-nationally comparable measures of employment concentration.

The NUTS classification is a hierarchical system for dividing up the economic territory of the European Union (EU). NUTS level zero corresponds to the country level. Increasing numbers indicate increasing levels of subnational disaggregation. The most disaggregated observed spatial units for which employment data are available for all sample countries over this period are NUTS-2 units. The EU defines NUTS-2 units as "basic regions for the application of regional policies."[26] The aim is to ensure that regions of comparable size all appear at the same NUTS level.[27] EU regulation defines minimum and maximum population thresholds for NUTS-2 regions. In Belgium, for example, the NUTS-2 units correspond to the country's eleven provinces.[28] In Austria, the NUTS-2 units correspond with the country's nine states or *Bundesländer*.

Despite population thresholds, NUTS-2 regions differ somewhat in terms of size, which generates a "unit" problem or what economic geographers refer to as, the "modifiable areal unit problem" (MAUP). This problem refers to the possibility that the results of statistical analysis of data for spatial zones can vary as a result of changing the zonal boundaries (Arbia 2001). In other words, using different levels of disaggregation can produce different results. To address this problem, I use NUTS-2 regions for all countries and years in my sample. Additionally, I condition physical space by the distribution of aggregate activity, which helps to minimize the modifiable areal unit problem.

---

[26] http://ec.europa.eu/eurostat/web/nuts/overview.
[27] http://ec.europa.eu/eurostat/web/nuts/principles-and-characteristics.
[28] http://ec.europa.eu/eurostat/web/nuts/national-structures-eu.

In the entropy index equation, y_{si} refers to employment in sector s in NUTS-2 unit i. Employment data come from Cambridge Econometrics.[29] For topographical concentration, $\bar{y}_s$ equals employment in sector s summed across all NUTS-2 units and divided by N. If sector employment is evenly distributed across N, the value of GE(1) will equal zero. For relative concentration, I calculate the percentage of employees in the manufacturing sector for each NUTS 2 unit. I then take the average across all NUTS-2 units in a given country. This value equals $\bar{y}_s$ for relative concentration calculations. The measure captures the degree of a sector's employment concentration relative to the geographic distribution of aggregate employment. The concentration variable ranges from zero to one with higher values indicating more geographic concentration.

MEASURING ELECTORAL INSTITUTIONS

I classify countries as being "PR" if proportional electoral rules apply to most of the seats in the lower house. More precisely, the variable *PR* equals 1 if proportional electoral rules are used to select most of the seats in the lower house and zero if most of the seats are filled via plurality. The data used to construct this variable come from the World Bank's Database of Political Institutions (Beck et al. 2001). As a robustness check, I also use an alternative data source to identify countries' electoral institutions (Golder 2005). Fortunately, Golder (2005) and Beck et al. (2001) are in complete agreement about the electoral institutions used in my sample countries.[30]

Although Germany has a mixed-member proportional electoral system, I code it as being PR. Mixed-member electoral systems, like Germany's, typically combine nominal-tier elections with list-tier elections (Shugart and Wattenberg 2001, Thames and Edwards 2006). In the former, citizens vote for individual candidates who accrue votes independently of party affiliation. In the latter, the distribution of legislative seats is according to votes for multiple candidates nominated on party lists. Germany's system is characterized as a mixed-member proportional system (MMP) by Shugart and Wattenberg because the total number of legislative seats received by a party is proportional to its list-tier results (Shugart and Wattenberg 2001). Since linking the tiers obtains outcomes that are proportional, MMP systems, like Germany's,

---

[29] For further information about the construction of the entropy indices, see Brülhart and Traeger (2005).
[30] There is one exception: Golder (2005) codes the French legislative elections in 1986 as being held via PR. This single difference does not change the results reported here.

often resemble pure PR systems. In fact, several previous studies of the effects of mixed-member systems have demonstrated the similarity between MMP and PR systems (e.g. Moser 2001, Cox and Schoppa 2002, Ferrara and Herron 2005, Thames and Edwards 2006). I therefore include Germany together with the pure PR systems in my sample.

Several additional measures of electoral institutions are used to test the robustness of the results. First, Gallagher's (1991) least-squares index, which measures the disproportionality between the distributions of votes and seats, is used.[31] This variable (*Disproportionality*) ranges, in theory, from zero to 100. Lower values indicate less disproportionality (Gallagher 1991). An electoral system where the legislature perfectly matches the distribution of votes would receive a score of zero. A legislature scoring 100 would consist only of individuals for whom not a single member of the electorate voted. Usefully, Gallagher's disproportionality index provides greater cross-national variation than the dichotomous measure of electoral systems (PR).

Second, the variable *Ballot*, measures parties' control over access to the ballot (Johnson and Wallack 2012). Access to the ballot determines, in part, the extent to which an electoral system is candidate or party centered. Politicians competing in candidate-centered systems tend to have geographically defined constituencies and thus an incentive to cater to geographically concentrated interests. In contrast, politicians competing in party-centered systems tend not to have geographically defined constituencies. Instead, candidates' electoral fortunes are determined by their party's national electoral success. Thus, politicians in party-centered systems have fewer incentives to cater to geographically defined interests.

Third, *Mean District Magnitude* refers to the number of representatives elected per district (Johnson and Wallack 2012).[32] Many arguments linking electoral institutions to policy outcomes focus on the importance of district magnitude. Indeed, many political scientists, including Duverger (1964), argue that district magnitude is the single-most important dimension by which electoral systems differ. Districts with only one representative, frequently referred to as single member districts

---

[31] This measure is calculated for each country-election-year. For nonelection years, the least-squares index is from the most recent previous election. If two elections occur in the same year, the average of the least-squares index (LSq) for that country-year is used.
[32] District size is distinct from district magnitude – although they are often conflated.

or SMDs, allow voters to assign credit to politicians for providing particularistic economic policies, such as subsidies. In multimember districts, voters observe the total amount of economic benefits provided to the district but not the amount produced by individual legislators. As a result, voters do not know which of their representatives to credit with providing particularistic policies (Ashworth and Bueno de Mesquita 2006). This reduces the electoral benefits of providing policies, such as subsidies, in multimember districts. If many legislators can claim responsibility for a policy with local benefits, each individual's incentive to provide such policies decreases (Lancaster 1986). Politicians in multimember districts may therefore provide fewer narrowly targeted economic policies than politicians in single-member districts (Magee, Brock, and Young 1989).

Some evidence exists to support this logic. Democracies with single-member districts face more allegations of providing narrowly beneficial trade protection at the World Trade Organization than democracies with multimember districts, holding all else constant (Rickard 2010). The number of GATT/WTO disputes filed against democracies with single-member districts is 186 percent higher, on average, then democracies with multimember districts (Rickard 2010). Moving from a multimember district system with an average of seven seats per district to a single-member district system increases the probability of being named as a defendant in a GATT/WTO dispute by more than six percentage points in a given year (Rickard 2010). This evidence suggests that district magnitude impacts economic policy. District magnitude may also have an indirect effect on economic policy by mediating the influence of electoral formulas (Carey and Shugart 1995, Carey and Hix 2011).[33] I examine this possibility in Chapter 6. In my sample, mean district magnitude ranges from one to 150. The variable *Mean District Magnitude* is logged to minimize the potential impact of outliers.

---

[33] For example, the effects of a proportional electoral formula may depend on the district magnitude. When district magnitude is high, electoral systems are relatively more proportional because smaller parties are more likely to win seats (Rae 1967, Taagepera and Shugart 1989, Cox 1997). A party would need to win more than 25 percent of the vote to guarantee a seat in a three-seat district but it would need to win only a little more than 10 percent of the vote to guarantee winning seat in a nine-seat district. The electoral system is likely to be disproportional whenever the district magnitude is small, irrespective of the particular formula used to translate votes into seats. For example, when PR is used in very small districts, as in Chile for example, its effects become similar to those of plurality elections.

EMPIRICAL MODEL

To investigate whether the effect of electoral systems on subsidies is conditional on the geographic distribution of voters with a shared interest in subsidies, the estimated models include an interaction term equal to the product of electoral systems and *Concentration*, along with both constitutive terms. More precisely, a partial-adjustment regression is estimated by ordinary least squares (OLS) with the following form and robust standard errors:

$$\text{Subsidies}_{it} = \beta_0 + \beta_1 PR_{it-1} + \beta_2 \text{Concentration}_{it-1} + \beta_3 PR_{it-1} {}^*\text{Concentration}_{it-1} + \beta X_{it-1} + \lambda_t + \varepsilon_{it}$$

where λ_t is a year fixed effect, and ε_{it} is an error term. The coefficient on β_3 is expected to be negatively signed. In other words, politicians in plurality systems will become relatively more responsive to demands for manufacturing subsidies as the geographic concentration of manufacturing employment increases.

Control Variables

X_{it-1} refers to a vector of control variables, which are lagged by one year to minimize concerns about endogeneity and take into account the fact that government budgets generally go through the legislative process and are voted on prior to the year in which spending occurs (Bawn and Rosenbluth 2006). All estimated models include three key control variables. The first is a measure of trade openness. Since manufacturing subsidies assist domestic producers in competing with lower cost foreign imports, countries more open to trade may spend more on subsidies. This is problematic if trade openness systematically relates to electoral institutions. Rogowski (1987) argued that countries dependent on international trade are more likely to have proportional electoral rules. To minimize the potential for a spurious correlation, a variable measuring trade openness as the sum of imports plus exports divided by GDP is included as a control.

A second important control variable is country size, measured by the natural log of a country's land area in square kilometers. Large countries will tend to have bigger manufacturing industries, which may increase government spending on manufacturing subsidies. Country size may also relate systematically to electoral systems; larger countries are more likely to have plurality electoral rules (Blais and Massicotte 1997). Controlling for country size minimizes the possibility of finding a spurious correlation between electoral rules and manufacturing subsidies.

The third control variable included in all estimated models is *GDP per capita*. Electoral support from lower-income voters may be relatively cheaper to "buy" using subsidies (Lindbeck and Weibull 1987, Dixit and Londregan 1996). Manufacturing subsidies may, therefore, be higher in countries whose voters have lower incomes as a result of strategic vote-maximizing spending by national governments.

Despite the potential relationship among the three control variables, standard tests show acceptable levels of multicolinearity.[34] Introduction of additional control variables one-at-a-time further minimizes multicolinearity. These additional control variables are:

Federalism, a dichotomous variable coded 1 for federal systems and zero otherwise. This is a potentially important control since the spending data refer only to central government expenditures. Data on general government spending, including that from local and regional governments, are often missing, and when available it tends to be less reliable than central government spending data (Persson and Tabellini 2003). Furthermore, the precise definition of local and regional governments' outlays are often not comparable among countries and time periods (Persson and Tabellini 2003). Given this, I use subsidy spending data for central governments.

Central government expenditures on subsidies may be lower in federal systems than nonfederal systems if some of the burden of subsidizing industries falls to regional and local governments. This would be particularly problematic if federal systems co-vary with electoral systems. In other words, if plurality electoral systems occur more often in federal systems, identifying a spurious negative correlation between plurality electoral rules and subsidy spending may be possible. To minimize this possibility, *Federalism* is introduced as a control variable.

Sector Employment equals the number of people employed in the manufacturing sector as a percentage of the total labor force. This is a potentially important control variable because the number of people employed in manufacturing may influence both the amount spent on manufacturing subsidies and the geographic distribution of manufacturing employees.

Left is a dichotomous variable coded one if the largest governmental party is left of center and zero otherwise. In general, governments' industrial policies tend to have only a minimal ideological component (Verdier 1995, McGillivray 2004, Rickard 2012c). Given this, the effect

---

[34] The variance inflation factor (VIF) is less than four for all variables included in the estimated models, as recommended by Huber, Ragin, and Stephens (1993).

of a government's ideology on total subsidies to the manufacturing sector is unclear. However, controlling for ideology is important because left-leaning governments tend to be associated with proportional electoral systems (Iversen and Soskice 2006, Rodden, unpublished manuscript). Failure to control for the ideological tendency of a government could result in mistakenly assigning explanatory power to electoral rules rather than ideology.

Concentration (squared) tests for the possibility that maximum political influence occurs at some intermediate level of geographic concentration. The existing literature on interest group politics in plurality systems hypothesizes a positive coefficient for the nonsquared concentration term and a negative coefficient for the squared term (Grier, Munger, and Roberts 1994). In other words, concentration may increase a group's political influence only up to some point. Beyond that point, any additional increase in geographic concentration may reduce the group's political influence.

EMPIRICAL RESULTS

Table 4.1 reports the coefficient estimates for the OLS regression of manufacturing subsidies on *PR, Concentration*, the key interaction term, and the control variables.[35] The coefficient estimates provide evidence that the geographic distribution of voters with a shared narrow interest matters for accurately specifying the effects of electoral rules. Subsidies for the manufacturing sector constitute a larger share of government expenditures in plurality systems than in PR systems when manufacturing employment is geographically concentrated. However, when employment is geographically diffuse, governments in PR systems assign relatively more of their budgets to subsidies than governments in plurality systems, holding all else equal.

The key interaction term, which equals the product of *PR* and *Concentration*, is included in columns 2 through 8. As a result, the estimated coefficients for *PR* in columns 2 through 8 report the marginal effect of *PR* for the unique case when *Concentration* equals zero (Brambor, Clark, and Golder 2006, Kam and Franzese 2007).[36] When *Concentration* equals zero, the effect of proportional electoral rules on

---

[35] See also Rickard (2012b).
[36] The unconditional (or average) effect of proportional electoral rules (PR) on manufacturing subsidies is positive and statistically significant as reported by the coefficient for *PR* in column 1 of Table 4.2. This finding may be interpreted as being consistent with results reported by Chang et al. (2010). They show that proportional systems are associated with higher consumer prices. If subsidies increase consumer

Table 4.1 *Effects of institutions and geography on subsidy budget shares*

	(1)	(2)	(3)	(4)	(5)	(6)	(7)	(8)
L.PR	0.300**	1.547***	1.548***	1.962***	1.421***	2.169***	1.032**	1.578***
	(0.118)	(0.480)	(0.480)	(0.463)	(0.469)	(0.472)	(0.481)	(0.504)
L.PR*L.Concentration		−40.72***	−40.72***	−47.90***	−34.91**	−55.89***	−26.05*	−43.64***
		(14.646)	(14.683)	(12.989)	(14.528)	(13.792)	(14.357)	(15.593)
L.Concentration	−15.61***	24.63*	24.63*	26.13**	18.34	32.18**	19.34	15.04
	(2.69)	(14.31)	(14.36)	(11.82)	(14.333)	(12.96)	(14.28)	(15.11)
L.Trade	0.017***	0.017***	0.017***	0.016***	0.017***	0.016***	0.010**	0.019***
	(0.003)	(0.003)	(0.003)	(0.004)	(0.003)	(0.004)	(0.004)	(0.003)
L.GDP per capita (log)	−1.863***	−1.875***	−1.875***	−1.797***	−1.846***	−2.008***	−0.937***	−1.740***
	(0.196)	(0.195)	(0.193)	(0.191)	(0.190)	(0.194)	(0.278)	(0.203)
L.Area (log)	0.259***	0.272***	0.272***	0.277***	0.288***	0.302***	0.181***	0.314***
	(0.057)	(0.057)	(0.062)	(0.071)	(0.060)	(0.084)	(0.068)	(0.065)
L.Federal			−0.000					
			(0.130)					
L.Employment				−0.479				
				(2.064)				
L.Left government					−0.168*			
					(0.099)			
L.Number of government parties						0.083		
						(0.071)		
L.Labor Mobility							5.965	
							(4.032)	

(continued)

L.Concentration^2						134.71***
						(50.325)
Constant	14.43***	13.17***	13.17***	13.37***	13.08***	14.71***	6.51**	11.41***
	(2.287)	(2.322)	(2.336)	(2.736)	(2.295)	(2.540)	(2.788)	(2.621)
Observations	227	227	227	169	227	169	209	227
R-squared	0.46	0.46	0.46	0.49	0.47	0.55	0.18	0.48

Robust standard errors appear in parentheses. All models include year fixed effects. Year coefficients are not reported due to space constraints.
*** $p < 0.01$, ** $p < 0.05$, * $p < 0.1$.

manufacturing subsidies is positive and statistically significant. In other words, governments elected via PR spend more of their budgets on subsides than governments in plurality systems when the geographic diffusion of manufacturing employees is exactly proportional to total employment.

The marginal effect of *PR* is positive and significant whenever *Concentration* is less than 0.033. When *Concentration* is less than 0.033, as it is for 67 percent of the sample, governments in proportional rule systems assign relatively more of their budgets to manufacturing-sector subsidies than governments in plurality systems, holding all else equal. Arguably, this is because manufacturing subsidies provide greater electoral benefits to politicians in PR systems than politicians in plurality systems when voters employed in the sector are geographically diffuse. If the geographic diffusion of manufacturing employees is exactly proportional to total employment (i.e. *Concentration* equals zero), politicians cannot use manufacturing subsidies to target voters in select electoral districts. In proportional systems where voters' geographic locations are relatively unimportant for the electoral success of parties, this is not a problem. However, in plurality systems where elections are won district-by-district, parties and politicians seek to target benefits to voters in geographically defined electoral districts.[37] Given this, subsidies are relatively less valuable to politicians competing for office in plurality systems when the beneficiaries of subsidies are geographically diffuse. Therefore, manufacturing subsidies account for a smaller share of government expenditures in plurality systems than in proportional systems when manufacturing employment is diffuse.

As expected, the positive marginal effect of *PR* on subsidies declines and eventually becomes negative as *Concentration* increases. The marginal effect of *PR* on subsidy budget shares is calculated across the observed range of *Concentration* using the relevant elements from column 2 in Table 4.1.[38] Figure 4.1 graphically reports these results. The solid line in Figure 4.1 represents the marginal effect of *PR* on

---

prices, the unconditional positive effect of PR on subsidies reported here is consistent with Chang et al. (2010).

[37] A large debate exists over precisely which electoral districts are most likely to be targeted by parties competing in plurality systems. See, for example, Cox and McCubbins (1986) and Dixit and Londregan (1996). Yet, all theories agree that parties and politicians in plurality systems want to target benefits to geographically defined constituents.

[38] The coefficient matrix and the variance-covariance matrix from column 2 are used to calculate the marginal effects of PR. For a complete description of these matrixes and the precise formulas used to calculate the marginal effects and standard errors, see Brambor et al. (2006). See also Rickard (2012b).

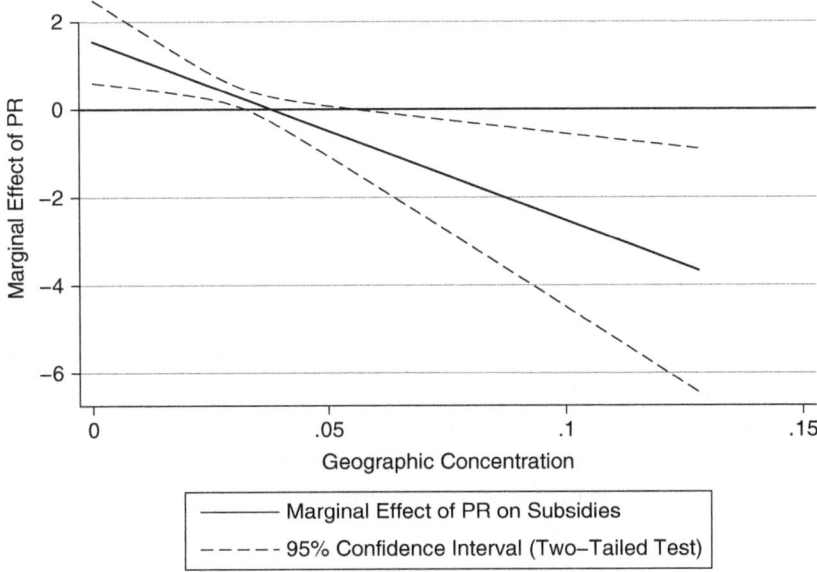

Figure 4.1 Marginal effect of proportional representation (PR) on subsidy budget shares

subsidy budget shares. The broken lines represent the 95 percent confidence intervals for two-tailed tests. Whenever both the upper and lower bounds of the confidence interval appear above (or below) the zero line, the marginal effect of PR is statistically significant at the 0.05 level (Brambor et al., 2006: 76).

Subsidies constitute a larger share of government expenditures in plurality systems than in proportional systems when voters with an economic interest in subsidies are geographically concentrated. When *Concentration* is greater than 0.054, proportional electoral rules have a negative marginal effect on the share of government spending devoted to subsidies. The reductive effect of plurality electoral rules on subsidy spending shares holds for nearly 15 percent of the sample.

At intermediate levels of concentration, no statistically significant difference exists between PR systems and plurality systems. The marginal effect of proportional electoral rules is not statistically different from zero when *Concentration* falls between 0.033 and 0.054. When manufacturing employment is neither concentrated nor diffuse, governments elected via different electoral systems allocate similar percentages of their budgets to manufacturing-sector subsidies, all else equal. In other words, there are some conditions under which electoral rules *do not* matter for governments' subsidy spending. This is an important finding – it suggests that existing theories about the economic effects of institutions must be revised in order to account for economic geography. Failing to account for economic geography generates incomplete, and at times inaccurate, conclusions about the effects of electoral rules on economic policy.

A few words about the control variables are in order. Countries more open to foreign trade spend relatively more on manufacturing subsidies, all else equal. This suggests that governments fund subsidies at least in part to shield domestic manufacturers from the effects of international trade (Rickard 2012b). Typically, the assumption in much of the literature on globalization and spending is that governments respond to trade openness by increasing spending on social welfare programs (e.g. Garrett 2001, Rudra 2002). However, the results reported here suggest that governments use multiple fiscal policies to offset the costs of trade, including subsidies. Understanding when and under what circumstances governments choose a particular fiscal policy in response to globalization is an important question for future research (Rickard 2012b).

Country size, measured by the natural log of a country's land area in square kilometers, is consistently positive and significant. Apparently, larger countries spend relatively more on subsidies for their manufacturing sectors as a share of total government expenditures (minus interest payments), all else equal.

GDP per capita has a negative effect on subsidies. A possible explanation is that richer voters have greater abilities to self-insure against price volatility and income risk. Alternatively, lower-income voters may be less costly to attract with subsidies (Dixit and Londregan 1996). Consequently, governments in countries with lower levels of GDP per capita spend more on subsidies.

A number of sensitivity analyses evaluate the robustness of the current study's key findings.[39] First, several different measures of electoral institutions are used to test the robustness of the results to alternative specifications of countries' institutions. Gallagher's (1991) disproportionality index is substituted for the dichotomous variable, *PR*. Results found using Gallagher's disproportionality index show that more proportional electoral systems favor geographically diffuse interests, while less proportional systems favor more concentrated interests. These results are reported in Table 4.2.

---

[39] For example, excluding the United Kingdom from the sample does not change the key results. Similarly, the key results are robust to alternative model specifications including OLS models with Driscoll-Kraay standard errors and the Newey and West estimator with lag length one.

Table 4.2 Effects of various features of countries' electoral systems on subsidy budget shares

	(1)	(2)	(3)	(4)	(5)	(6)
L.Disproportionality	-0.001	-0.033*				
	(0.011)	(0.017)				
L.Disproportionality*L.Concentration		1.058**				
		(0.442)				
L.Ballot			-0.298***	-0.999***		
			(0.070)	(0.202)		
L.Ballot*L.Concentration				19.493***		
				(5.388)		
L.MDM					-0.122**	0.315
					(0.0527)	(0.229)
L.MDM*L.Concentration						-14.40**
						(7.001)
L.Concentration	-15.12***	-23.56***	-23.21***	-36.71***	-18.64***	7.952
	(2.740)	(4.960)	(3.740)	(5.861)	(3.395)	(13.99)
L.Trade	0.018***	0.021***	0.015***	0.017***	0.017***	0.021***
	(0.003)	(0.004)	(0.004)	(0.004)	(0.004)	(0.005)
L.GDP per capita (log)	-1.879***	-1.801***	-1.988***	-2.215***	-1.849***	-1.993***
	(0.202)	(0.203)	(0.195)	(0.202)	(0.183)	(0.199)

(continued)

Table 4.2 (continued)

	(1)	(2)	(3)	(4)	(5)	(6)
L.Area (log)	0.231***	0.287***	0.212***	0.272***	0.0995	0.163*
	(0.057)	(0.069)	(0.063)	(0.057)	(0.087)	(0.092)
Constant	15.144***	13.746***	18.070***	19.978***	17.07***	16.90***
	(2.374)	(2.588)	(2.353)	(2.376)	(2.458)	(2.505)
Observations	227	227	194	194	180	180
R-squared	0.44	0.46	0.51	0.54	0.493	0.506

Robust standard errors appear in parentheses. All models include year fixed effects. Year coefficients are not reported due to space constraints. *** $p < 0.01$, ** $p < 0.05$, * $p < 0.1$.

Empirical Results

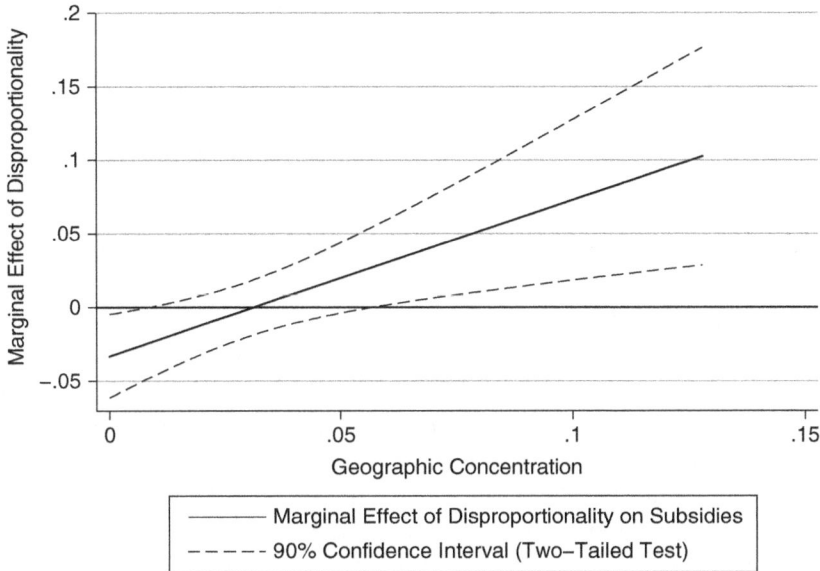

Figure 4.2 Marginal effect of disproportionality on subsidy budget shares

Using the relevant elements of the variance-covariance matrix from column 2 in Table 4.2, the marginal effect of *Disproportionality* is calculated across the entire range of *Concentration*. The marginal effect of disproportionality is displayed graphically in Figure 4.2. When beneficiaries are geographically diffuse, the marginal effect of disproportionality is negative. More precisely, the marginal effect of *Disproportionality* is negative and statistically significant at the 90 percent level of confidence for a two-tailed test when *Concentration* is less than 0.009.[40] This result suggests that when beneficiaries are geographically diffuse, governments in more proportional systems, such as Austria, tend to spend more on subsidies than governments in less proportional systems, such as Spain. More precisely, an increase in the disproportionality of electoral outcomes decreases subsidy spending shares when manufacturing employment is geographically diffuse. As employment becomes more and more concentrated, the negative marginal effect of *Disproportionality* decreases in magnitude and eventually becomes positive and statistically significant. When *Concentration* is greater than 0.055, an increase in disproportionality increases government spending on subsidies. In other words, governments elected via less proportional systems tend to spend more on subsidies for

---

[40] This holds for approximately 10 percent of the sample.

concentrated groups than governments elected via more proportional systems, all else equal. These results corroborate those found using the simple dichotomous measure of electoral rules. More proportional systems tend to spend more on geographically diffuse groups, all else equal.

Table 4.2 also reports the estimated coefficients on *Ballot* and *Mean District Magnitude*. The estimated coefficients on *Ballot* demonstrate that geographically concentrated sectors win more subsidies in candidate-centered systems than in party-centered systems. Geographically diffuse sectors win more subsidies in party-centered systems than in candidate-centered systems. Arguably, this is because relevant constituencies are not geographically defined in party-centered systems. Candidates maximize their chances of being in office by working to increase the party's share of the national vote. In contrast, politicians' best electoral strategy in candidate-centered systems is to appeal directly to their geographically defined constituents.

The estimated coefficients on *Mean District Magnitude* show that governments elected from multimember districts allocate less government money to subsidies than governments elected from single-member districts when manufacturing employment is geographically concentrated. More precisely, when *Concentration* is greater than 0.27, *Mean District Magnitude* has a negative and significant marginal effect on subsidy budget shares, all else equal. The magnitude of the reductive effect of *Mean District Magnitude* increases as *Concentration* increases. Interestingly, MDM does not have a significant effect when looking only at PR systems.[41]

In sum, different measures of electoral institutions report strikingly consistent results. Taken together, these results demonstrate that economic geography is a necessary consideration needed to accurately specify the effects of electoral rules on policy outcomes.

The OLS models estimated thus far do not account for the fact that the choice of electoral institutions is unlikely to be random (Boix 1999). However, it seems improbable that the choice of electoral rules would be endogenous to manufacturing subsidies in the short- to medium term. Similarly, it seems unlikely that electoral rules would be endogenous to changes in the geographic concentration of manufacturing employment in recent decades. In fact, there is no significant difference between the average levels of geographic concentration in PR systems and plurality systems.[42] The sample mean value of *Concentration* is equal to 0.035 in

---

[41] See Chapter 6.
[42] This may be because no direct relationship exists between geographic concentration and electoral systems. Instead, as Cusack et al. (2007) argue, the effects of geographic concentration on electoral rules may be conditional on the type of asset investment.

PR countries and 0.031 in plurality countries. The difference (−0.004) is not statistically significant, as demonstrated by a two-sample t-test with equal variances. The fact that the geographic concentration of manufacturing employment is no higher in plurality systems than in PR systems, on average, helps to minimize concerns about the potential endogeneity of electoral institutions.

To further allay concerns about endogeneity, I estimate a two-stage least-squares model. Following Persson and Tabellini (2003), Evans (2009), and others, I used indicators of the historical periods during which a country's current electoral rules were adopted as instruments in the first stage of the model. The distribution of current electoral rules vary with the age of the rules (Persson and Tabellini 2003). Experience of other democracies and prevalent political and judicial doctrines shift systematically over time and may explain why the distribution of current electoral rules vary with the age of the rules. To exploit this temporal pattern, I construct three dummy variables that correspond to the periods 1921–1950, 1951–1980, and post-1981, which take a value of 1 if the current electoral rule originated in the respective period, and zero otherwise.[43] Although countries' electoral systems are associated with the year in which countries' constitutions were adopted, the date at which a country adopts its constitution is unlikely to directly affect manufacturing subsidies.

The results from the second-stage of the 2SLS model are reported in Table 4.3. As before, the marginal effect of *PR* on subsidies is positive and statistically significant at low levels of geographic concentration. As *Concentration* increases, the positive marginal effect of *PR* declines and eventually becomes negative. At high level of *Concentration*, the marginal effect of *PR* is negative, substantively large, and statistically significant. In sum, the key results are robust to an alternative model specification that relaxes the assumption that electoral systems are exogenous.[44]

Before concluding, I briefly discuss possible alternative explanations for the reported results. Production factors employed in geographically concentrated sectors may confront higher adjustment costs than factors in geographically diffuse sectors. This raises the possibility that asset specificity rather than *Concentration*, per se, explains the reported results. To test for this possibility, a measure of labor mobility is introduced as

---

[43] Persson and Tabellini (2003) demonstrate that these particular time periods best describe the pattern of electoral system adaptation.
[44] If anything, correcting for potential endogeneity appears to reduce the standard errors on the estimated marginal effect of *PR*.

Table 4.3 Second stage results of the effects of PR on subsidy budget shares

	(1)	(2)	(3)	(4)	(5)	(6)	(7)	(8)
L.PR	0.562***	3.834***	3.929***	3.633***	3.755***	3.991***	3.343***	4.000***
	(0.179)	(0.579)	(0.575)	(0.569)	(0.587)	(0.541)	(0.582)	(0.600)
L.PR*L.Concentration		−107.78***	−111.86***	−95.80***	−103.96***	−108.99***	−94.02***	−114.94***
		(19.933)	(19.981)	(17.581)	(20.617)	(17.343)	(19.384)	(20.957)
L.Concentration	−19.98***	86.56***	89.46***	68.24***	82.21***	78.82***	80.80***	78.80***
	(2.720)	(19.326)	(19.381)	(15.729)	(20.209)	(16.268)	(18.719)	(20.221)
L.Trade	0.014***	0.015***	0.015***	0.013***	0.014***	0.013***	0.009*	0.016***
	(0.003)	(0.003)	(0.003)	(0.004)	(0.003)	(0.004)	(0.004)	(0.003)
L.GDP per capita (log)	−1.935***	−1.981***	−2.067***	−1.881***	−1.961***	−2.124***	−1.271***	−1.823***
	(0.171)	(0.171)	(0.175)	(0.176)	(0.168)	(0.166)	(0.270)	(0.186)
L.Area (log)	0.317***	0.338***	0.405***	0.341***	0.354***	0.391***	0.272***	0.389***
	(0.053)	(0.056)	(0.061)	(0.067)	(0.058)	(0.083)	(0.063)	(0.062)
L.Federal			0.419***					
			(0.122)					
L.Employment				0.091				
				(1.964)				
L.Left government					−0.138			
					(0.102)			
L.Number of government parties						0.141**		
						(0.069)		

(continued)

L.Labor Mobility						1.130
						(3.735)
L.Concentration						158.86***
						(48.54)
Constant	15.46***	12.41***	12.30***	12.24***	12.25***	13.27***	6.88***	10.17***
	(1.88)	(1.96)	(2.05)	(2.55)	(1.95)	(2.26)	(2.47)	(2.33)
Observations	203	203	203	155	203	155	185	203
R-squared	0.53	0.51	0.54	0.54	0.52	0.61	0.22	0.53

Notes: From 2SLS Model. Robust standard errors appear in parentheses. All models include year fixed effects. Year coefficients are not reported due to space constraints. *** $p < 0.01$, ** $p < 0.05$, * $p < 0.1$. In the first stage model, *PR* is predicted using historical periods, as described in the text.

a control variable. This measure estimates the adjustment costs facing workers in the manufacturing sector by calculating the rate of labor movement between industries in the sector (Wacziarg and Wallack 2004). The rate of movement varies according to the costs to workers of voluntarily entering and exiting different industries. Higher rates of movement indicate lower adjustment costs. Including *Labor Mobility* in the estimated model does not change the key results and suggests that geographic concentration is, in fact, the mechanism linking economic geography to policy outcomes rather than adjustment costs.

Another plausible alternative explanation is the number of parties in government. I argue that electoral systems affect politicians' incentives to cater to certain constituencies and that these incentives influence spending on subsidies. Alternatively, electoral systems may influence subsidy spending via the number of parties in government. Single-party governments are most common in plurality systems, while PR systems are more likely to foster multiparty governments. Multiparty governments tend to spend more money than single-party governments because multiparty governments negotiate less efficient logrolls (Bawn and Rosenbluth 2006). This raises the possibility that multiparty governments will spend more on subsidies than single-party governments. If this is the case, the reported results may not be the consequence of the electoral dynamics or reelection incentives but rather the accountability and bargaining dynamics induced by multi- versus single-party governments.

To test for this possibility, I included the number of parties in government as an additional control variable. The addition of this control variable does not change the key results. Electoral systems affect subsidy spending independent of their effects on the composition of government.

CONCLUSION

Economic geography interacts with electoral institutions to shape government policy. Governments elected via proportional rules spend more of their budget on subsidies to geographically diffuse groups as compared to governments elected via plurality rules. In contrast, governments elected via plurality rules spend more of their budget on subsidies to geographically concentrated groups as compared to governments elected via PR, all else equal. In the following chapter, I supplement the large-N quantitative results reported here with qualitative evidence from two cases that illustrate the political mechanisms linking geography and institutions to economic policy.

5

The Power of Producers

Successful Demands for State Aid

Why do democratically elected governments spend more money on subsidies in some countries than in others? One reason may be the structure of a country's economy or the particular products a country produces. If a country's economic structure shapes its industrial policy, only a fool would try to compare countries' industrial policies in their entirety. Indeed, policy-makers I interviewed expressed skepticism about the wisdom of comparing countries' industrial policies.[1] Even comparing industrial policy across two relatively similar countries, such as Norway and Sweden, raised eyebrows among policy-makers.[2] While academics tend to be more comfortable making generalizations and cross-national comparisons, I take seriously the potential challenges of comparing countries' industrial policies in their totality. For this reason, I focus exclusively on subsidies to the manufacturing sector in the previous chapter. In this chapter, I focus exclusively on subsides for a single industry. Focusing on one industry allows for a meaningful comparison of government support for business between countries.

In this chapter, I examine government subsidies for the wine industry. I compare subsidies for winemakers in two countries: France and Austria. The wine industries in these two countries, while not identical, share many similar features making them ideal for a comparative case study. The wine industry has the added advantage of having largely "exogenous" geographic patterns. Only certain geographic locations are suited to wine production, and factors, such as weather, soil, and terrain, primarily determine winemakers' location decisions. The wine industry consequently provides a valuable case study with which to investigate the links between geography, institutions, and policy.

---

[1] In person interview with Innovation Norway staff members Pål Aslak Hungnes and Per Melchior Koch in Oslo, Norway on June 19, 2014.
[2] In person interview with Innovation Norway staff members Pål Aslak Hungnes and Per Melchior Koch in Oslo, Norway on June 19, 2014.

France and Austria provide a valuable comparative case study because their electoral institutions differ significantly. Austria's highly proportional system[3] provides an unambiguous contrast to France's majority-plurality system where candidates win office by obtaining a majority of votes in the first ballot or, failing that, a plurality of votes at the second ballot (Elgie 2005). In Austria, candidates fill a party's legislative seats in the order in which they are listed on a party's list and legislative seats are awarded to parties based on their vote shares. Given these electoral institutions, I expect to see different types of government assistance for producers. In France, I expect to find economic benefits geographically targeted to select groups. In contrast, I expect economic benefits to be more broadly beneficial in Austria.

Using a detailed case-selection criterion described below, I identify two subsidy programs that I investigate in detail – one in each country. Both programs conform to the expectations derived from my theory. The French subsidy program exclusively benefits producers that are geographically concentrated in a small area of the country. In contrast, the Austrian program assists producers across the entire country – irrespective of their geographic location. Both subsidy programs ran afoul of the European Union's rules on state aid. Rules on state aid seek to ensure a level playing field for all companies across the union.

EU state aid rules introduce an important factor to the model of economic policy-making developed in Chapter 3: international politics. Up to this point, I have focused exclusively on domestic politics arguing that incumbents in different countries have varied electoral incentives to supply certain types of economic policies. Their incentives stem from a country's electoral institutions and economic geography. But domestic electoral incentives are only part of the policy-making story. National policy decisions are increasingly influenced by international agreements. International rules, like those of the European Union (EU) and World Trade Organization (WTO), regulate countries' economic policies. As a result, governments frequently must consider the international implications of their economic policy decisions. Some policy choices will conflict with governments' international obligations. In these cases, governments must weigh the costs of violating international rules against the domestic benefits of implementing the "illegal" policy. For the subsidies examined here, the governments in both Austria and France concluded that the benefits gained at home from providing the subsidy outweighed the international costs of violating EU state aid rules.

---

[3] Austria's electoral system produces some of the most proportional results in the world (Müller, 2005: 397).

The domestic benefits of these subsidies were large precisely because of the constellation of economic geography and electoral institutions. Together, electoral institutions and economic geography robustly predict the likelihood that a government will violate EU state aid rules – illustrating that domestic politics shape not only national economic policy but also international economic relations.

CASE SELECTION

Myriad subsidy programs exist. Given the ubiquity of government subsidies, it would be far too easy to cherry pick cases that best fit my theory. To guard against this, I use a methodical, multistep selection criterion. First, I identify EU member-states' subsidy programs using documents published by the European Commission.[4] Every year, the Commission compiles a list of member-states' subsidies. The Commission monitors states' subsidies to ensure that they comply with EU rules that regulate subsidies (or "state aid" in EU-nomenclature). These rules seek to ensure a level playing field for all companies across the union, and for this reason member-states are generally not allowed to subsidize their own producers. EU rules also aim to prevent subsidies from being used as an alternative form of protection in the absence of traditional barriers to trade, such as tariffs. EU rules prohibit subsidies that distort competition or have negative effects on intra-EU trade (Besley and Seabright 1999).

EU rules further require states to inform the Commission of their subsidy programs (Besley and Seabright 1999). Each notification triggers a preliminary investigation by the Commission into whether or not the subsidy is compatible with EU regulations. In this way, the Commission acts more like a police patrol than a fire alarm (McCubbins and Schwartz 1984).

Not all subsidies are prohibited by EU rules.[5] Subsidies can be used in a number of circumstances because EU treaties recognize that government

---

[4] I focus on subsidies that actually transpired (rather than those that did not). While this may be considered by some to be equivalent to "selecting on the dependent variable," it is appropriate here given my goal. I seek to illustrate the political mechanisms that link geography, electoral institutions, and policy. For this reason, I focus on cases where the outcome of interest (i.e. subsidies) actually occurred. Additionally, as a practical matter, it is extraordinarily difficult to identify subsidies that did not happen. Most subsidy programs fail to materialize long before they get to the legislature and as a result they are virtually invisible to researchers ex post.

[5] Articles 87(2) and, more importantly in practice, 87(3) of the EC Treaty provide exceptions, thereby leaving states room for maneuver regarding the use of state aid (Blauberger 2009).

aid is sometimes necessary. Subsidies are permissible, for example, if they promote the execution of an important project of "common European interest" or remedy a serious disturbance in the economy of a member state. Aid is also permitted to facilitate the development of certain economic activities where such aid does not adversely affect trading conditions to an extent contrary to the common interest.[6] Subsidies that aim to improve firm productivity and economic growth by removing or reducing market failures (or system failures) are also compatible with the rules of the Single Market (Stöllinger and Holzner 2016).

Given the varied circumstances under which subsidies are allowed under EU state aid rules, governments can assist producers using subsidies while fully complying with EU rules. In fact, most government-funded subsidies comply with EU rules (European Commission 1995). In 1997, the Commission registered 657 subsidies and found that 98 percent of them complied with EU rules.[7] Given this, the universe of EU compliant subsidies is too large to usefully identify cases for in-depth study. Therefore, as a first step, I limit my sample to subsidies formally investigated by the European Commission. If doubts exist as to the compatibility of a subsidy with EU state aid rules, the Commission opens a formal investigation. At the end of the investigation, the Commission makes a final ruling. If the Commission issues a negative decision, the measure is deemed to be incompatible with EU rules. I limit my sample to cases where the Commission made a negative or partially negative decision under Article 93(21) of the EC Treaty with or without recovery. In 2014, only 26 out of 593 subsidy investigations resulted in a "negative decision with recovery" in which the Commission ruled that the aid was noncompliant with State Aid rules and required the government to "recover" the aid. If the Commission makes a negative decision "with recovery," the Commission requires the member state to recover the aid already paid out with interest from the beneficiaries.[8]

---

[6] EU subsidy rules are regularly reviewed in response to demands from member-states and the European Council. The Council called for fewer but better targeted subsidies to boost the European economy following the 2008 global economic crises and in May 2014, the Commission adopted revised subsidy rules.

[7] In sectors other than agriculture, fisheries, transport, and coal (European Commission 1997).

[8] The Commission subsequently opens a "recovery case" to enforce the implementation of its decision. On June 30, 2014, there were forty-nine active pending recovery cases. (Author's calculations using data available from the European Commission's database of Competition Cases available at http://ec.europa.eu/competition/elojade/isef/index.cfm?pa=2). If the member state does not comply with the decision in due time, the Commission may refer it to the European Court of Justice (ECJ).

Case Selection

The aim of recovery is to remove the undue advantage granted to companies.

While this selection criterion usefully limits the sample size, it may introduce bias. Subsidies incompatible with EU rules may differ from subsidies that comply with state aid rules. Noncompliant subsidies may, for example, stem from particularly intense domestic pressures for assistance. Leaders facing intense demands for aid may be more willing to provide subsidies in violation of EU state aid rules. On one hand, such dynamics might generate an upward bias in the apparent level of domestic demand for subsidies. On the other hand, intense domestic pressure for subsidies may help to make the mechanisms linking electoral institutions and economic geography to policy more visible to observers ex post.

In fact, the politics behind noncompliant subsidies appear to be similar to the politics that lead to EU-compliant subsidies. The same variables that predict subsidy spending (i.e. geographic concentration and electoral institutions) also explain the cross-national variation in non-EU compliant subsidies.[9] To demonstrate this, I identify all negative or partially negative decisions by the European Commission with respect to manufacturing-sector subsidies under Article 93(21) of the EC Treaty (with or without recovery) from 1988 to 2000. Using these original data, I estimate a negative binominal regression model of non-EU compliant subsidies on PR, geographic concentration, the key interaction term, and important control variables. The results are reported in Table 5.1.

Governments elected via proportionality are more likely to violate EU state aid rules than governments elected via plurality when the beneficiaries are geographically diffuse. When manufacturing employment is geographically diffuse (i.e. when *Concentration* equals zero), the expected number of "illegal" manufacturing subsidies is

---

[9] No evidence exists to suggest that the Commission's rulings on state aid are biased (Besley and Seabright 1999). All illegal subsidy programs are equally likely to be ruled against by the Commission. In fact, some conclude that the Commission's negative decisions on state aid are "as-if" random (Besley and Seabright 1999). Some scholars suggest that the idea of systematic bias in the commission's rules is simply "implausible" (Franchino and Mainenti, 2016: 8). The EU's strongest member (Germany) is ruled against most frequently by the Commission during this period. The fact suggests that the Commission's decisions are not biased in favor of the union's most powerful states. But if any bias does exist, it would arguably make it more difficult to find support for my argument. If, for example, illegal subsidy programs in some member-states are more likely to be ruled against than those in other states, the pattern of negative decisions would deviate from the pattern I predict based on domestic political considerations. Any bias that does exist in the Commission's decisions would therefore make it more rather than less difficult to find support for my argument.

Table 5.1 *Effects of PR on non-EU compliant subsidies*

	(1)	(2)	(3)	(4)
PR	2.049***	2.092***	2.078***	1.872***
	(0.551)	(0.585)	(0.554)	(0.651)
Concentration	30.546***	30.899***	30.090***	28.833***
	(7.210)	(7.308)	(6.915)	(7.761)
PR*Concentration	−18.980**	−19.726*	−19.169**	−16.867*
	(9.484)	(10.195)	(9.685)	(10.423)
GDP (log)	1.433***	1.394***	1.404***	1.377***
	(0.164)	(0.181)	(0.163)	(0.209)
Area (log)		0.053		
		(0.190)		
Economic Growth			−0.089	
			(0.079)	
Employment				0.018
				(0.035)
Alpha (log)	−0.767*	−0.779*	−0.707*	−0.708*
	(0.418)	(0.437)	(0.410)	(0.411)
Constant	−41.77***	−41.37***	−40.73***	−40.67***
	(4.653)	(4.444)	(4.658)	(5.404)
Observations	156	156	156	156

Notes: Robust standard errors in parentheses; *** $p < 0.01$, ** $p < 0.05$, * $p < 0.1$; EU country-years only; manufacturing sector only; results are robust to one-year lags.

higher in PR countries than plurality countries, as demonstrated by the positive and statistically significant coefficient on *PR* in Table 5.1. The positive coefficient is statistically significant across all four models.

Figure 5.1 graphically illustrates the results reported in Table 5.1. In Figure 5.1, the coefficients from column 4 of Table 5.1 are transformed to indicate the predicted number of noncompliant subsidies per country-year. The predicted numbers of subsidies that violate EU state aid rules in a given country-year is low, ranging from zero to five. This result accords with the fact that the European Commission typically rules against only a handful of subsidies in any given country for any given year.

When *Concentration* is less than 0.07, the expected number of noncompliant subsidies is higher in PR countries than plurality countries. In other words, governments elected via proportionality fund more noncompliant subsidies than governments elected via plurality when the beneficiaries are geographically diffuse. This finding is consistent with the argument that incumbents in plurality systems have few incentives to

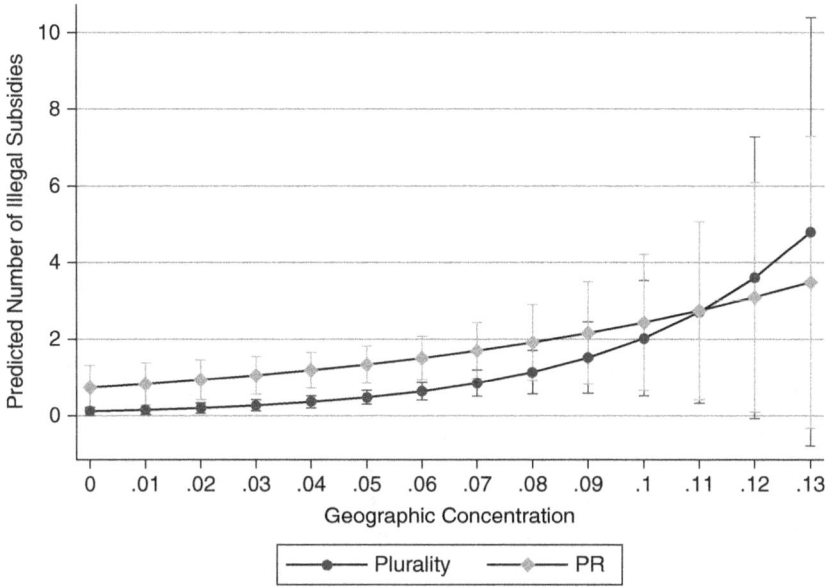

Figure 5.1 Predicted number of non-EU compliant subsidies (bars indicate 95 percent confidence interval)

fund subsidies for groups spread across multiple districts. Funding subsidies for groups across multiple districts is an inefficient reelection strategy for incumbents in plurality systems. Instead, incumbents in plurality systems seek to fund programs that exclusively benefit groups concentrated in their own district because doing so is an efficient way to maximize their chances of reelection. In contrast, incumbent parties in PR systems seek to increase their share of the national vote by subsidizing groups spread across the country. Leaders in PR countries are therefore more willing to fund subsidies for diffuse groups than leaders in plurality countries – even in violation of EU state aid rules.

As *Concentration* increases, the number of noncompliant subsidies increases in both PR and plurality countries. However, the rate of increase is greater in plurality countries, as compared to PR countries. When *Concentration* is higher than 0.11, the predicted number of noncompliant subsidies is larger in plurality countries than PR countries. However, the 95 percent confidence intervals overlap. At high levels of geographic concentration, the predicted number of noncompliant subsidies does not differ meaningfully between PR and plurality countries.

The same factors that explain the variation in subsidy spending between countries (i.e. geographic concentration and electoral rules) also explain the cross-national variation in noncompliance with EU state aid rules. These results suggest that the politics behind noncompliant subsidies are similar to those of compliant subsidies. The findings reported in Table 5.1 also help to minimize concerns about possible bias in the population of noncompliant subsidies. Previous studies have similarly found no evidence of bias in the Commission's rulings on state aid (Besley and Seabright 1999, Franchino and Mainenti, 2016: 8). Illegal subsidy programs appear to be equally likely to be ruled against by the Commission regardless of the member-state in which they originate. The Commission's decisions do not appear to be biased in favor of the union's most powerful states. In fact, the strongest EU member-state, Germany, is ruled against frequently by the Commission. Some scholars even suggest that the Commission's negative decisions on state aid are "as-if" random (Besley and Seabright 1999). If any bias does exist in the Commissions' decisions, it would arguably make it more difficult to find support for my argument. If illegal subsidy programs in some member-states are more likely to be ruled against than those in other states, the pattern of negative decisions would deviate from the pattern I predict based on domestic political considerations. Any bias that does exist in the Commission's decisions would therefore make it more difficult to find support for my argument.

Although only a small proportion of state subsidies run afoul of EU rules, the sample of negative or partially negative decisions remains too large to usefully identify cases for an in-depth investigation of the political dynamics behind subsidies. Therefore, as a further step, I limit my sample to negative or partially negative decisions made during the period from 1990 to 2000. I focus on these years because EU state aid rules – and the Commissions' enforcement of them – remained relatively constant over this period.[10] This consistency ensures that I compare like cases.

However, even this sample contains too many cases for a meaningful in-depth study. Therefore, as a final selection criterion, I limit my sample to a single industry. A single-industry case study ensures that the myriad factors that vary between different industries are held constant. Declining industries, for example, tend to win more state aid than prosperous industries. A single industry study holds such factors constant and consequently helps to isolate the effects of electoral institutions and economic geography. In this chapter, I focus exclusively on subsidies to the wine industry.[11]

---

[10] Even though the formal instruments of state aid control were laid out in the 1957 Treaty of Rome, it was only in the mid-1980s that enforcement became a Commission priority. Important changes in state aid policy and enforcement occurred in the year 1990 and again in 1999 (Cini 2001, Blauberger 2009).

[11] Cognac is technically a type of brandy rather than a wine per se. However, like wine, it is produced from grapes and its production methods must meet certain legal requirements. Cognac is made by distilling wine, and then aging the resulting spirit

Why Wine?

The wine industry provides a particularly useful case study because the geography of wine production is largely exogenous to policy. Factors such as weather and soil are far more important determinates of the location of vineyards than politics. Vines must be planted in areas where the topography and climate are conducive to grape growth. And producers often operate on or near vineyards to minimize the time between harvest and pressing. In Europe, wine is frequently produced at the vineyard in order to qualify for the prestigious geographic classifications. Often, a wine's denomination can only be attributed if the grapes are grown and pressed in the delimited region (Meloni and Swinnen 2013). In sum, the geography of wine production is largely exogenous to politics.

Because the geographic pattern of wine production is generally exogenous to politics, concerns about the direction of causation – or what causes what – are minimized. In theory, producers could strategically choose their geographic location to take advantage of a country's political system. Producers in plurality systems, for example, could tactically locate themselves in concentrated groups to maximize their chances of winning particularistic economic policies, like subsidies. In this way, economic geography could be a function of a country's electoral rules. However, average employment concentration levels are nearly identical in plurality systems and PR systems, which suggests that producers do not strategically locate themselves in order to take advantage of the electoral system. Producers in plurality systems are just as concentrated as producers in PR systems, on average. This finding is consistent with the large and growing literature on firms' location decisions, which shows that factors other than government policy are far more important to firms' location decisions.[12]

The wine industry illustrates one reason why economic geography looks broadly similar in plurality and PR systems: producers' location options are limited by their production requirements. Wine producers cannot choose from an infinite number of locations to set up shop. Winemakers must be near high quality grape-producing vineyards, and vineyards must be located in areas with suitable climate, soil, and terrain. They have limited location options and as a result they have little ability to strategically locate their production processes to take advantage of the electoral system. Because the geography of wine production is determined

---

in wood barrels. The Cognac label can only be applied to the spirit if it was produced in a specific geographic region of France.
[12] European Commission (2003).

primarily by factors other than countries' electoral institutions, government aid to the wine industry provides a useful case with which to examine the impact of electoral institutions and geography on economic policy.

The Outcome of the Selection Process

My case selection strategy has several additional benefits. First, only countries bound by identical international rules are compared with one another. The EU rules regulating the wine industry are highly restrictive and as a result, it would be misleading to compare wine subsidies in EU countries to wine subsidies in non-EU countries. France, for example, is constrained by EU subsidy rules but South Africa faces no such restrictions. Comparing government aid to winemakers in France to subsidies in South Africa would be misleading. Usefully, my sample is restricted to EU member countries, which are bound by identical international rules.

Second, my empirical strategy ensures that I did not selectively pick cases to fit my theory. The sample under investigation encompasses the entire population of illegal wine subsidies during the period from 1990 to 2000. During this period, there are three wine subsidy programs that violate EU state aid rules – one each in France, Austria, and Germany. I examine the French and Austrian programs in detail in this chapter. I set aside the German subsidy because of the complexities of Germany's mixed electoral system. The mixed-member proportional system makes it more difficult to isolate the precise mechanisms linking geography, institutions, and policy. However, the German case broadly fits my theory. Leaders elected from single-member districts provided a subsidy to a geographically concentrated group of winemakers located on just 982 hectares (3.8 square miles) in a single federal state, an area that constituted less than 0.1 percent of the total area in Rhineland-Palatinate. In sum, the universe of illegal wine subsidies in my sample period conform to my theoretical expectations. Before turning to the details of the two illegal wine subsidies, I begin first with brief, general discussion of wine subsidies.

WINE SUBSIDIES

Many governments subsidize wine production. Various policies assist winemakers including subsidized loans, minimum price schemes, drought relief schemes, water subsidies, aid for vine-pulling or "grubbing," and tax breaks. While government assistance for wine producers is common and widespread, EU rules regulate member-states' support for the wine industry.

Of all the wine consumed around the world today, 60 percent is subject to EU rules (OIV 2007). While the EU has generic subsidy rules (or "state aid" in EU nomenclature), the wine industry is subject to special regulations.[13] Aid to wine producers is allowed, for example, if the objective is to reduce wine production or improve quality (without leading to increased production).[14] Governments can provide subsidies to offset the costs of converting vineyards from one varietal to another (but they must not increase production potential). Similarly, governments can fund programs that aim to reduce the production of wines for which there is no market (via conversion). Subsidies are also allowed under EU rules if they improve the quality of the product without leading to increased production.[15] In sum, governments interested in assisting wine producers must navigate a complex web of EU rules and regulations.

At times, governments run afoul of the EU rules. In many cases, governments appear to have inadvertently violated EU regulations in their attempts to assist domestic winemakers. In the following case studies, I identify where the demands for subsidies came from and explore why the demands were successful in winning support from the government – even in violation of EU state aid rules. I begin with the case of France.

SUBSIDIES FOR FRENCH COGNAC PRODUCERS

The French subsidy program granted supplemental financial aid to wine producers in the demarcated Cognac region. The designated Cognac region, bordered by the Atlantic Ocean on the west and the medieval city of Angoulême on the east, covers the département of Charente-

---

[13] In the EU today, wine production is regulated and controlled by a comprehensive Wine Common Market Organization in the framework of the Common Agricultural Policy.

[14] These rules prohibit, for example, government aid for the planting of winegrowing areas from September 1, 1988, except where such aid contains criteria to ensure, in particular, that the objective of reducing production or improving quality is achieved without leading to increased production (32001D0052 2001/52/EC: Commission Decision of September 20, 2000 on the State aid implemented by France in the winegrowing sector (notified under document number C(2000) 2754 Official Journal L 017, 19/01/2001 P. 0030 – 0037 available at http://eur-lex.europa .eu/legal-content/EN/TXT/?uri=CELEX:32001D0052.

[15] These subsidies are permissible under the WTO's 1994 Uruguay Round Agreement on Agriculture (URAA) because they are considered "blue box" subsidies.

Maritime and most of Charente. The French government approved two decrees to grant aid for winegrowing holdings in the Charente-Maritime and Charente area on March 12, 1999 and April 6, 2000, for the 1998/1999 and 1999/2000 wine years, respectively.

Of all the grapes grown in Charente, 95 percent are destined for Cognac distillation (Yagoda 2009). Cognac is technically a type of brandy rather than a wine per se. However, like wine, it is produced from grapes and its production methods must meet certain legal requirements. Cognac is made by distilling wine, and then aging the resulting spirit in wood barrels. The Cognac label can only be applied to the spirit if it was produced in a specific geographic region – the eponymous Cognac region of western France.

Four measures were agreed by the French government to help Cognac producers in Charente. These programs included: (1) supplemental aid for improving the vine population; (2) technical support for producers; (3) the promotion of Cognac; (4) supplements to the grubbing premium – a financial incentive to voluntarily pull up grape vines. "Grubbing up" subsidies are intended to reduce supply and subsequently raise prices. Producers in the region qualified for a supplemental payment of €1,524 per hectare to help convert vineyards traditionally used to produce Cognac, to the production of other wines known as "local wines."[16] The estimated budget for the subsidy was FRF 10,000,000 (€1,524,000).[17] Only producers located in the Cognac demarcated region were eligible for this aid.

Government support for winemakers in France

The Cognac subsidy is just one example of state support in a country with a long history of government assistance for winemakers. French governments of various ideological persuasions have assisted wine producers over the decades.[18] In 1953, for example, the "Code du Vin" reestablished subsidies to uproot vines, as well as subsidies for surplus

---

[16] JORF n°85 du 11 avril 1999 page 5387, https://www.legifrance.gouv.fr/affichTexte.do?cidTexte=JORFTEXT000000211372.

[17] JORF n°85 du 11 avril 1999 page 5387.

[18] French wine producers benefit from EU assistance. In fact, France is the third largest beneficiary of EU subsidies for the wine industry (i.e. of all aid paid by the EAGGF guarantee behind the wine industry). France falls behind Spain and Italy with 19 percent of the aid (Ministère de l'agriculture et de la pêche Mise à jour: janvier 2006 Direction des affaires financières et de la logistique Le vin et les produits de la viticulture http://agriculture.gouv.fr/IMG/pdf/vin_2004.pdf). EU payments to the wine industry represented 2.3 percent of total CAP expenditure in 2009, according to the latest European Commission figures available. The largest producer nations,

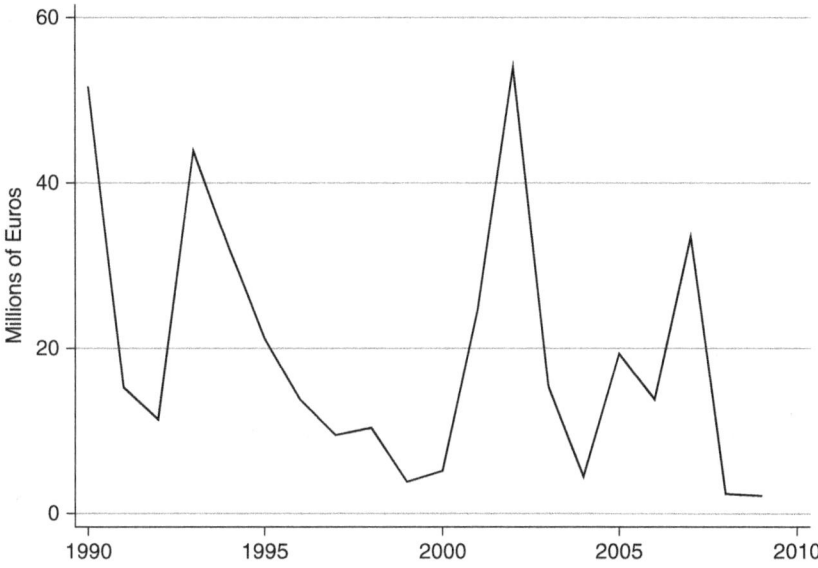

Figure 5.2 Total national government financial aid to the wine industry, in millions of euros. Source: Public support to agriculture (*Les concours publics à l'agriculture*).

storage and compulsory distillation (Meloni and Swinnen 2013). In more recent years, government support for the industry has fluctuated, as illustrated in Figure 5.2. Between 1990 and 2004, the level of public assistance to the French wine industry varied.[19] The two largest spikes in government assistance occurred in election years, which suggest politics may play a role in state support for the industry. Elected leaders may seek the electoral support of industry-specific interest groups in the run up to election contests.

Interest Group Politics in France

Historically, no single interest group represented the entire wine industry in France. Instead, pressure groups were organized around Appellations, or geographically defined areas (Meloni and Swinnen 2013). In an attempt to regulate quality, the French government created an explicit link between the "quality" of the wine and its production region (the *terroir*). The regional

France, Spain, and Italy, took the majority of the subsidies, which totaled EUR 1.31 billion Euros.

[19] These variations are due, in part, to market fluctuations, including the occurrence of overproduction (Ministère de l'agriculture et de la pêche Mise à jour: janvier 2006 Direction des affaires financières et de la logistique Le vin et les produits de la viticulture, http://agriculture.gouv.fr/IMG/pdf/vin_2004.pdf).

boundaries of Bordeaux, Cognac, Armagnac, and Champagne wines were established between 1908 and 1912. These regional boundaries were referred to as *Appellations*. In 1935, a law formalized the *Appellations d'Origine Contrôlées* (AOC). Producers and winegrowers coalesced into AOC groups. For example, in the Champagne AOC region, three powerful lobbying groups emerged: the *Fédération des Syndicats de la Champagne* that represented winegrowers; the *Syndicat du Commerce des Vins de Champagne* that promoted exports of the *Maisons de Champagne;* and the *Association Viticole Champenoise* that stood for the interests of both winegrowers and Maisons.[20] The geographically defined AOC associations became powerful interest groups whose members typically lived and worked in relatively concentrated geographic locations. These interest groups lobby the government for economic assistance and have been quite successful in winning subsidies (Meloni and Swinnen 2013). The political influence of these groups is due, in part, to the country's electoral institutions.

Electoral Institutions in France

France's electoral institutions make it politically expedient for elected leaders to respond to the demands of geographically concentrated interest groups. The French electoral system is classified as being a two-ballot majority-plurality system (Elgie 2005). A candidate is elected by winning either a majority of votes in the first ballot or failing that, a plurality of votes at the second ballot, hence the term "majority-plurality" (Elgie, 2005: 122). A candidate is elected in the first round if she obtains an absolute majority of the total votes cast, provided this amount is equal to a quarter of registered voters in a given constituency. If no candidate obtains an absolute majority in the first round, a second round election is held. To be a candidate in the second round, a candidate must have obtained 12.5 percent of the registered voters in the first round. At the second ballot, a relative majority of cast votes is sufficient to get elected.[21]

France's electoral system incentivizes candidates to focus tenaciously on the demands of their geographically defined constituents (i.e. the voters located in the geographically defined constituency in which the candidate is seeking election). Only these voters matter for a candidate's (re)election

---

[20] The creation of these associations was made possible, in part, by the 1884 French law that legalized labor unions (Simpson 2011).
[21] In the case of an equal number of votes cast between the two candidates, the elder candidate is elected.

chances. Voters located in other districts are irrelevant. Therefore, to maximize their chances of holding office, politicians seek to target economic benefits to producers geographically concentrated in their own district. In this way, the two-ballot system has similar effects as a simple majoritarian or plurality system (Elgie 2005).

France's electoral institutions generate incentives for politicians to geographically target economic benefits to their own constituents. A survey of sitting French MPs revealed that 41.2 percent agreed it was their duty to "represent above all his/her constituency and his/her region" (Brouard et al. 2013).[22] As one French MP put it: "An MP is a gardener. He has a big garden, his constituency, and he has to go to Paris in order to get fertiliser."[23] Subsidies are one example of such fertilizer. Subsidies benefit producers and when producers are geographically concentrated in a MP's district, they have incentives to secure such programs. In the case of French wine producers, geographically concentrated producers share common economic interests, and these shared interests form the basis for organized interest groups, which lobby legislators for targeted economic benefits.

Adding further to the incentives for localism is the fact that national legislators in France tend to have close ties to local constituencies (Costa and Kerrouche 2009). A long-standing tradition exists that candidates elected to the National Assembly have a local base (if not they must be "parachuted" into a position of responsibility at the local level as soon as possible [Elgie, 2005: 131]).[24] In the 1997–2002 legislature, for example, 97 percent of all deputies simultaneously held some sort of elected office at the local level (Elgie, 2005: 131). All three of the legislators that lobbied for the wine subsidy held local positions in addition to their national legislative seat. For example, Marie-Line Reynaud served as a member of the Municipal Council of Jarnac. Jarnac is home to several Cognac distilleries, including Courvoisier. Didier Quentin served on both the Charente-Maritime's General Council and the Poitou-Charentes Regional Council. Dominique Bussereau also served as a regional councilor for Poitou-Charentes and a general councillor for the Charente-Maritime département.

Given this tradition, French legislators may pay relatively more attention to the concerns of their local constituency than their counterparts elsewhere, even controlling for electoral rules (Elgie, 2005: 131).

---

[22] Brouard et al. (2013).
[23] Quoted in Brouard et al. (2013: 146). This quote came from a member of the centre-right political party Union for a Popular Movement (*Union pour un mouvement populaire*/UMP).
[24] Note however that this is not due to the electoral system per se.

However, constituents' demands succeed at the national level because of France's national electoral institutions. Although France has three tiers of subnational government, it is a unitary, centralized state, in which most authority is exercised by a power base in Paris (OECD 2010). The centralized power is controlled by leaders elected to office via a majority-plurality system, which incentivizes the provision of narrowly targeted policies, such as the wine subsidy I examine here.

In sum, France's electoral institutions align with the geography of the French wine industry to produce geographically targeted support programs. The financial support provided to producers in the Cognac region is just one such example.

The Politics Behind Targeted Support for Cognac

Cognac producers faced difficult economic conditions in the late 1990s. Cognac sales began falling as markets in Asia, most notably Japan, suffered. Asia was one of the key markets for French brandy and the economic crisis in Asia hit the French Cognac industry hard.[25] The decline in global demand resulted in an accumulation of cognac stock, which in turn drove down cognac prices and subsequently profits. Many smaller producers were pushed to the verge of bankruptcy. Some growers even destroyed a segment of their vines in an attempt to sustain prices.[26]

In September 1998, winemakers in the region took to the streets to demand government assistance.[27] Over 700 viticulturists and grape farmers set up barriers around the cities of Cognac and Jarnac to demand subsidies and tax breaks in compensation for declining sales of French brandy.[28] Local grape growers and cognac producers were joined by members of the agriculture union, Movement for the Defense of Family Farmers (MODEF), in a massive tire burning campaign.[29] The protests caused serious disruption to the area. The four-day blockade ended when the government promised targeted financial assistance. Growers were promised aid to cut overproduction, as well as cheap-rate loans to tide them over until they had been paid for their harvests. Full details of the

---

[25] *Dans les Charentes, la crise du cognac met à mal toute la region*, Le Monde, November 26, 1997; Reuters, Oct 1, 1998 via Factiva accessed on Nov 3, 2014.
[26] Reuters, Oct 1, 1998 via Factiva accessed on Nov 3, 2014.
[27] Bussereau. Parliamentary question no 468 Question published in the OJ: 05/10/1998 page: 5339 Source: http://questions.assemblee-nationale.fr/q11/11-468QOSD.htm
[28] Reuters, Oct 6, 1998 via Factiva accessed on Nov. 3, 2014
[29] Torula News 1998 Global Torula News Archive Sept 29, 1998 www.cognacnet.com/torulanews/torulanews0998.htm.

assistance program were to be worked out with the Agriculture Ministry.[30]

Ultimately, four measures were agreed by the French government to help Cognac producers including a "grubbing up" subsidy – a financial incentive to voluntarily pull up grape vines. Only producers located in the Cognac demarcated region were eligible for this financial assistance.[31] The geographically targeted nature of this subsidy conforms to my theoretical expectations. In France, where candidates compete in single-member districts under a majority-plurality system, I anticipate that subsidies will tend to be targeted to geographically concentrated groups, like the Cognac producers of Charente. Politicians competing for office in single-member districts have incentives to supply economic rents to their geographically defined constituents in order to maximize their chances of (re)election. Subsidies provide one policy tool by which to achieve this goal. It is no surprise then that this subsidy program benefits exclusively to narrow, geographically concentrated producer group.

Parliamentary Questions

Legislators' electoral strategies influence their legislative behavior (Ames 1995, Carey and Shugart 1995). Specifically, I argue that electoral institutions shape legislators' reelection strategies, which, in turn, influence their legislative behavior. One important form of legislative behavior is parliamentary questions. Parliamentary questions (PQs) give legislators an opportunity to pose questions to government ministers either in the legislative chamber or via writing. PQs provide a unique insight into parliamentarians' concerns because they are a form of legislative behavior typically not organized by political parties (Martin 2011a). Additionally, the number of questions allowed from the floor is limited, and legislators consequently treat PQs as a scarce resource. They carefully consider what policies to agitate for and which issues to raise with ministers in their parliamentary questions.

I identify all of the PQs asked in the French legislature that relate to the Cognac subsidy program.[32] During the period from 1997 to 2000, three

---

[30] Reuters, Oct 1, 1998 via Factiva accessed on Nov 3, 2014.
[31] The estimated budget was FRF 10,000,000 (EUR 1,524,000) and the aid was to be granted in the form of a payment of FRF 10,000/ha (EUR 1,524) to winegrowers (JORF n°85 du 11 avril 1999 page 5387).
[32] The legislative record provides valuable evidence of legislators' priorities and policy preferences. However, legislators may use multiple means to maximize their chances of winning office. Legislators may, for example, lobby ministers directly for key policies. Such lobbying may not appear in the legislative record. Legislators may

legislators asked eleven parliamentary questions seeking subsidies for wine makers in the Cognac region. All of the PQs were addressed to the Minister of Agriculture and Fisheries. Details of these questions can be found in Table 5.2.

Although each of the three legislators came from a different political party, they all represented the same region. Marie-Line Reynaud, a member of the governing Socialist (*Socialiste*) party, came from the *Charente Département* (*circonscription* 2). Didier Quentin was elected from the *Département Charente-Maritime* (*circonscription* 5) as a member of the center-right opposition party *Rassemblement pour la République*. Dominique Bussereau represented the *Département Charente-Maritime* (*circonscription* 4) as a member of *Démocratie libérale et indépendants*, which was a center-right opposition party. This pattern suggests that ideology and partisanship are relatively less important than geography for explaining legislators' policy priorities. Legislators from different parties, but the same region, demanded subsidies for their geographically concentrated constituents (i.e. Cognac producers). In effect, geography trumped ideology.

The questions, recorded in the legislative record, highlight the legislators' efforts to secure state aid for their winemaking constituents. Bussereau, for example, urged the Minister of Agriculture to provide targeted assistance to the winegrowers of the Cognac region on the floor of the National Assembly on October 5, 1998. He argued that the government should make an "unprecedented effort" because the region has very serious economic difficulties.[33] Bussereau described the winemakers' recent protests and urged the government to act before such demonstrations became violent. In his PQ, Bussereau took the opportunity to propose explicit support measures, including subsidized loans for grape growers, financial aid for distillation, and a minimum price target, which would work to "quickly improve farmers' incomes."[34]

Reynaud, a Socialist legislator from Charente, asked the most parliamentary questions related to the Cognac subsidy. In fact, she asked seven of the eleven PQs related to aid for Charente winemakers. On the floor of the National Assembly, Reynaud implored the Minister to subsidize Charente wine producers. Specifically, she called for supplemental aid for Charente area winemakers to support diversification – that is, the uprooting

lobby ministers behind closed doors. Given this, it is important to remember that the legislative record likely provides only part of the picture.

[33] Author's translation. Original text can be found at http://questions.assemblee-nationale.fr/q11/11-468QOSD.htm.

[34] Author's translation. Original text can be found at http://questions.assemblee-nationale.fr/q11/11-468QOSD.htm.

Table 5.2 *Parliamentary questions about the French Cognac subsidy*

Number	Legislator	Party	Département	Published in JO (day/month/year)	Minister's Reply (day/month/year)
11ème législature – QE 56084	Dominique Bussereau	Démocratie libérale (DL)	Charente-Maritime	25/12/2000	19/02/2001
11ème législature – QE 53674	Didier Quentin	Rassemblement pour la République (RPR)	Charente-Maritime	13/11/2000	05/02/2001
11ème législature – QE 32186	Marie-Line Reynaud	Socialiste (PS)	Charente	28/06/1999	18/10/1999
11ème législature – QE 17374	Didier Quentin	Rassemblement pour la République (RPR)	Charente-Maritime	27/07/1998	05/07/1999
11ème législature – QE 9823	Marie-Line Reynaud	Socialiste (PS)	Charente	09/02/1998	25/01/1999
11ème législature – QE 5366	Marie-Line Reynaud	Socialiste (PS)	Charente	27/10/1997	02/02/1998
11ème législature – QE 2380	Marie-Line Reynaud	Socialiste (PS)	Charente	25/08/1997	05/07/1999
11ème législature – QE 2379	Marie-Line Reynaud	Socialiste (PS)	Charente	25/08/1997	25/01/1999
11ème législature – QE 2378	Marie-Line Reynaud	Socialiste (PS)	Charente	25/08/1997	10/11/1997
11ème législature – QOSD 1257	Marie-Line Reynaud	Socialiste (PS)	Charente	25/12/2000	10/01/2001
11ème législature – QOSD 468	Dominique Bussereau	Démocratie libérale (DL)	Charente-Maritime	05/10/1998	07/10/1998

Notes: JO refers to the *Journal officiel de la République française*, which is the official gazette of the French Republic. It publishes the major legal official information from the national Government of France.

of vines for Cognac and replanting new varieties. She argued that this government assistance was necessary to lift "the Cognac region out of its structural crisis."[35]

The Minister of Agriculture came to the legislative chamber in person to answer these legislators' questions. Minister Le Pensec (Socialist) said that he agreed with the seriousness of the economic crisis facing wine growers in Charente and Charente-Maritime.[36] He stated that the situation had his "full attention" and that he sought to implement "a more ambitious structural adjustment program." He clarified that such an adjustment program "will require undoubtedly adequate resources." He specified that he wanted to provide short-term relief via distillation subsidies and grubbing up premiums. He expressed particular support for young winemakers, as well as those who make a greater effort of collective organization.

The Prime Minister ultimately issued a decree granting aid to wine makers in Charente-Maritime and Charente.[37] This example illustrates how a handful of engaged legislators can succeed in winning targeted aid for their constituents. In this case, the legislators did not need to convince a plurality of fellow legislators as the program was not passed by legislative degree. Instead, they needed to convince the Agriculture Minister and the Prime Minister to fund the subsidy.

The Role of Partisanship

In this case, both the Agriculture Minister and the Prime Minister were socialists. Is this merely a coincidence or does ideology play a role in governments' subsidy decisions? In France, governments of various ideological persuasions have assisted winemakers. Although the Cognac subsidy was provided by a Socialist Prime Minister, the legislators who demanded the subsidy were not, uniformly, Socialists. In fact, two of the three legislators requesting assistance for cognac producers came from right-leaning, opposition, parties. Bussereau, a member of the center-right Démocratie libérale (DL) party, explicitly minimized the role of partisanship in his demands for cognac subsidies. He clarified that his demands for subsidies had nothing to do with the fact that he was from an opposition party. Instead, his insisted that his demands reflected only the dire economic situation of Cognac growers in his constituency. Bussereau argued that demands for targeted assistance could reasonably come from

---

[35] Author's translation. Original text can be found at http://questions.assemblee-nationale.fr/q11/11-1257QOSD.htm.

[36] Author's translation. Original text can be found at http://questions.assemblee-nationale.fr/q11/11-468QOSD.htm.

[37] JORF n°85 du 11 Avril 1999 page 5387.

any of the areas' legislators. And, indeed, legislators from two other parties made similar demands on the floor of the National Assembly. The Minister of Agriculture, himself a Socialist, identified legislators from other parties who shared similar concerns regarding Cognac producers in his legislative speech.

In this case, electoral incentives appear to have outweighed partisan politics. The socialist government allocated financial help not just as a way of funneling pork to its own constituents or members of parliament. However, we cannot know for certain whether the outcome would have been different if the Prime Minister had not been a socialist. Thus, the French case does not fully resolve the question of how governments' ideology matters for subsidies.

The Role of Electoral Competitiveness

What role, if any, did the competitiveness of elections play in the Cognac subsidies? Among legislators from the region, those elected in safe seats did not press for subsidies for their winemaking constituents via parliamentary questions. Jean-Claude Beauchaud, for example, won a convincing victory in 1997 with 64.8 percent of the second round vote.[38] Beauchaud was from the same party as the Prime Minister and as such was in a unique position to win subsidies for his winemaking constituents. Yet, Beauchaud, did not ask for assistance for wine producers in the Cognac region.[39] Perhaps Beauchaud did not feel the need to lobby on behalf of his constituents given the lack of competition he faced in the 1997 election. Given his lack of constituency service, it comes as no surprise that Beauchaud's vote share declined by 5 percentage points to 59 percent in 2002. Although he was returned to the legislature in 2002, it was by a far less convincing margin.

---

[38] Elections were held for all the seats in the National Assembly following the premature dissolution of the legislature on April 22, 1997. In the 1997 election, the Socialist Party won 246 seats. It did not obtain an absolute majority (289 seats) and thus relied on the support of its allies to govern. These allies included the 37 Communists, 13 Radical Socialists, 16 various left, and eight Greens who entered the National Assembly for the first time. The National Front, instrumental in defeating candidates from the moderate right, won only a single seat. The RPR and its ally the Union for French Democracy (UDF), which together lost over 200 seats, had a total of 256 seats with the "various right." Taking heed of these results, President Chirac called on Socialist Lionel Jospin to form the government, which was appointed on June 4 and included Communist and Green Ministers. Jospin was a Socialist. France thus began a period of "cohabitation" between a rightist President and a leftist Government. The aid in question was enacted by decree by the Socialist Prime Minister (Source: JORF n°85 du 11 Avril 1999 page 5387).

[39] At least not in public.

Legislators from the Charente and Charente-Maritime region who faced more competitive elections in 1997 were more likely to press for subsidies in subsequent years. On average, legislators that asked for Cognac subsidies won their seats by smaller margins than the regional average. The average vote share for legislators elected from the nine *circonscriptions* in the region was 55.9 percent. The average vote share for legislators who asked for wine subsidies for the region was only 53.75 percent. Quentin, for example, won office with 55.37 percent of the second round vote and Reynaud won with 54.81 percent. Of the three legislators who ask for subsidies, Bussereau faced the most competitive race. In the second round of the 1997 election, Bussereau defeated his opponent by only 1,136 votes winning just 51.07 percent of the vote.

Bussereau's race was the most competitive, as illustrated in Table 5.3. Given how slim his margin of victory was in 1997, he had every incentive to try to win extra votes by speaking out in support of subsidies. Bussereau worked hard to win subsidies for Cognac makers. He submitted written questions and also asked parliamentary questions addressed to the Minister of Agriculture. In these questions, he argued vigorously for subsidies for the area's winemakers. He even faced off with the Minister on the floor of the National Assembly in an attempt to win subsidies for his wine-producing constituents. This case suggests that legislators elected in more competitive districts will tend to work harder on behalf of their constituents in single-member district, plurality systems.

In this case, the Prime Minister, Lionel Jospin (Socialist), did not have to choose between safe or competitive districts when awarding subsidies to Cognac producers. Because of the geography of the Cognac designated area, subsidies to Cognac producers went to both safe and competitive districts. The Charentes region is comprised of five departments, two of which contain Cognac wine producers: Charente and Charente-Maritime. For the Socialist party, Charente was a safe *département*. In the second round of the 1997 legislative elections, the Socialists won 57.22 percent of the vote in Charente. All four legislators from the département were Socialists. However, the Socialist Party faced stiffer competition in Charente-Maritime. The centrist UDF (*Union pour la Démocratie Française*) party won 36.45 percent of the vote in Charente-Maritime the second round of the 1997 legislative elections. In contrast, the Socialists won just 30.31 percent of the vote. Following the 1997 election, only two of the five legislators from Charente-Maritime were Socialists.

Given the geographic distribution of Cognac producers, the subsidy provided benefits to both safe and competitive districts. As a result, this

Table 5.3 *Electoral competitiveness in Charentes*

Département	Circonscription	Candidate 1	Party	Votes	% Votes	Candidate 2	Party	Votes	% Votes
CHARENTE	1	VIOLLET	SOC	22117	53.23	MOTTET	UDF	19433	46.77
CHARENTE	2	REYNAUD	SOC	22813	54.81	HOUSSIN	RPR	18808	45.19
CHARENTE	3	LAMBERT	SOC	28883	56.36	RICHEMONT DE	RPR	22364	43.64
CHARENTE	4	BEAUCHAUD	SOC	26465	64.8	RIFFAUD	UDF	14378	35.2
CHARENTE-MARITIME	1	CREPEAU	PRS	31495	59.57	CLERC	UDF	21377	40.43
CHARENTE-MARITIME	2	GRASSET	SOC	26791	53.07	BRANGER	UDF	23691	46.93
CHARENTE-MARITIME	3	ROUGER	SOC	28790	54.86	ROUX DE	UDF	23690	45.14
CHARENTE-MARITIME	4	CALLAUD	PRS	25981	48.93	BUSSEREAU	UDF	27117	51.07
CHARENTE-MARITIME	5	BILLOT ZELLER	SOC	24153	44.63	QUENTIN	RPR	29964	55.37

case cannot help to resolve the debate as to which districts economic rents, such as subsidies, disproportionality flow. While some argue that parties target benefits to safe districts to reward their core supporters, others believe that parties target benefits to swing districts in an attempt to win over nonpartisan or unattached voters. Given the geographic diffusion of Cognac producers, the Socialist Prime Minister did not have to choose between safe or competitive districts. Instead, he could target benefits to both safe districts and swing districts using a single program – precisely because of the geographic distribution of the relevant producers.

The Policy Outcome

Winemakers and grape growers articulated their economic interests, legislators lobbied on behalf of their constituents, the minister drafted a policy response, and the subsidy program was ultimately approved by the Prime Minister. The Cognac subsidy reflects the importance of economic geography and political institutions for policy outcomes. Producers with shared economic interests were geographically concentrated. Their economic fortunes rose and fell together, and in response to declining sales they lobbied for government assistance. France's political institutions incentivized the regions' legislators to respond to the demands of their geographically defined constituents. The regions' legislators worked to persuade the national government to provide financial assistance to the geographically concentrated Cognac producers. In sum, the French case illustrates the importance of geographic concentration for economic policy outcomes in plurality electoral systems.

SUBSIDIES FOR AUSTRIAN FARM-GATE WINE MERCHANTS

The Austrian subsidy program provides a nice contrast to the French program. In Austria, the government provided farm-gate wine merchants with a subsidy equal to 7.58 percent of the final product price.[40] Farm-gate wine merchants (*Ab-Hof-Verkauf soll*) sell wine directly to consumers on their farms. Farm-gate sales play an important role in the wine sector in Austria (Euromonitor International 2015). From 2009 to 2014, farm-gate sales accounted for 18 percent of total wine sales, on average (Euromonitor International 2015). Although farm-gate wine merchants have lost market share to supermarkets and food/drink/tobacco specialists over time, as

---

[40] http://eur-lex.europa.eu/legal-content/EN/TXT/?uri=CELEX:31999D0779.

Table 5.4 *Austrian farm-gate wine sales as a percentage of total sales*

2009	2010	2011	2012	2013	2014
18.6%	18.4%	18%	17.9%	17.8%	17.7%

Source: Euromonitor International (2015).

illustrated in Table 5.4, they continue to account for nearly one-fifth of the market (Euromonitor International 2015, Statistics Austria 2015).

In 2014, there were 11,602 wine producing holdings in Austria.[41] About 4,900 of these holdings registered their wines as PDO or "quality wine."[42] According to the Minister of Agriculture, this registration indicates the "classical" farm-gate wine merchants.[43] In other words, 42 percent of Austrian wine producers are farm-gate wine merchants. Farm-gate merchants are geographically dispersed across all of Austria's wine regions[44] and can consequently be characterized as a narrow, yet geographically diffuse interest group.

The Austrian government provided subsidies to farm-gate wine merchants via a tax break. Tax breaks are a type of government subsidy that allow producers to keep a larger share of their earnings.[45] Tax breaks reduce government revenues and therefore are regarded by the European Commission as "being granted through state resources." Because tax breaks advantage some producers over others and are granted through state resources, the European Commission considers tax breaks to be

---

[41] Farm-gate wine merchants are not statistically registered in Austria according to the Federal Ministry of Agriculture, Forestry, Environment and Water Management. Rudolf Schmid, Federal Ministry of Agriculture, Forestry, Environment and Water Management, Vienna, Austria. Email correspondence on December 23, 2015.
[42] Rudolf Schmid, Federal Ministry of Agriculture, Forestry, Environment and Water Management, Vienna, Austria. Email correspondence on December 23, 2015.
[43] Rudolf Schmid, Federal Ministry of Agriculture, Forestry, Environment and Water Management, Vienna, Austria. Email correspondence on December 23, 2015.
[44] Wine is produced in significant quantities in four of the nine federal states, including Lower Austria, Vienna, Burgerland, and Styria.
[45] Although financial subsidies require a monetary outlay, governments forgo revenues by providing tax breaks to producers. When a government provides a tax break, its budget is affected in much the same way as if it had spent money. Tax exemptions are equivalent to a waiver of the fiscal revenue that would otherwise have to be paid by the producer to the government. The exemption is borne by the public budget insofar as it reduces revenues and therefore is regarded as being granted through state resources (Steenblik 2012). However, tax breaks are not included in the measure of fiscal subsidies used in Chapter 4 because they do not appear on the expenditures side of governments' budgets.

a form of subsidy (i.e. state aid). The European Commission consequently regulates tax breaks using the same rules that regulate fiscal subsidies (Plender 2003, Blauberger 2009).

The Austrian government exempted farm-gate wine merchants from the country's beverage tax, which is levied at a rate of 7.58 percent of the retail price for sales of alcoholic beverages to consumers. The government exempted wine (heading numbers 220421 and 220429 of the Common Customs Tariff) and other fermented beverages (heading number 220600 of the Common Customs Tariff) where these products are sold directly to the consumer at the place of production. By doing so, the government provided geographically diffuse farm-gate wine merchants with an effective subsidy equal to 7.58 percent of the final product price.

Austria's subsidy program differed notably from France's program. In France, only producers in a small geographic area received economic assistance from the government. In contrast, Austria's tax break helped all farm-gate wine merchants – regardless of their geographic location. In fact, the European Commission confirmed in their investigation of the program that the farm-gate tax break did not apply exclusively to any particular region of Austria.[46] Austria's subsidy program provided geographically diffuse benefits while, in contrast, France's subsidy program provided geographically concentrated benefits. I argue that this difference is due, in part, to the countries' electoral institutions. In the following section, I briefly describe Austria's electoral system and its influence on policy outcomes.

Austria's Electoral Institutions

Austria provides a useful contrast to France because the countries' electoral institutions differ significantly. Austria's electoral system is a complex three-tier PR system that allocates seats in the National Council (*Nationalrat*) at the regional, state, and national levels via proportionality. The National Council is one of two houses in the Austrian Parliament and is sometimes referred to as the lower house. The constitution endows the National Council with significantly more power than the second chamber. For National Council elections, Austria is divided into nine regional electoral districts.[47] The nine regional districts are further subdivided into forty-three local electoral districts, which typically have more than one seat. The number of seats assigned to each local district depends on the district's population, as established by the

---

[46] http://eur-lex.europa.eu/legal-content/EN/TXT/?uri=CELEX:31999D0779.
[47] The nine regional electoral districts correspond to the nine states of Austria.

most recent census. Votes are cast first and foremost for a particular political party.[48] As a result, elections in Austria are party-centered rather than candidate-centered. Parties compete against one another in a near-perfect proportional system to maximize their share of votes and subsequently seats in the National Council.

The number of votes required to win a seat is the number of votes cast divided by the number of seats assigned to the district in question. For example, if 200,000 votes are cast in a five-seat local district, 40,000 votes are needed to win one seat. If a party wins 81,000 votes out of the 200,000 votes cast, it is entitled to two seats to be taken by the first two candidates on the party's local district list.[49] Since 80,000 votes would have been sufficient to win two seats, 1,000 votes are left unaccounted for by this first round of tallying. However, these "extra" votes are not lost to the party. Instead, they go to the regional level.[50] The system at the regional level is analogous to that used on the district level; the number of seats assigned to a regional district is simply the number of seats assigned to one of its constituent local districts but not filled during the first round of tallying. Any vote not accounted for on the regional level either is dealt with on the federal level, provided that the party it has been cast for has obtained at least 4 percent of the federal total vote. The D'Hondt method is used to allocate any National Council seats remaining to be filled. The system produces highly proportional results. In fact, within the universe of real-world PR systems, Austria's is one of the most fully proportional (Müller, 2005: 397).[51]

Prior to reforms in 1992, Austria was a fully closed-list system where voters confronted party lists at all levels – regional, state, and federal (Müller, 2005: 404). In other words, voters chose a party but could not express support for any individual candidate. Reforms in 1992, which became operational in the 1994 election, allowed voters to express some preferences for individual candidates.[52] However, in practice, voters do

[48] In addition to voting for a party list, voters may express a preference for one individual candidate. A candidate receiving sufficient personal votes can rise in rank on his or her party's district list. In theory then, voters have a certain degree of influence as to which particular individual wins which particular seat. However, in practice, this almost never happens.
[49] Political parties submit separate, ranked lists of candidates for each district, regional or local election, in which they have chosen to run. They also submit a federal-level list.
[50] More precisely, any vote not accounted for on the local level is dealt with on the regional level, provided that the party it has been cast for has obtained at least 4 percent of the regional total vote.
[51] However, a slight bias in favor of large parties persists (Müller, 2005: 414).
[52] Electors vote for a party and may cast preferential votes for one regional list candidate and one state list candidate.

not intervene strongly in the process of allocating seats to individual candidates (Müller, 2005: 409). As a result, parties have a de facto monopoly in determining the composition of parliament (Müller, 2005: 401). Parties allocate the legislative seats they win to the individual candidates of their choosing.

Because party organizations are virtually free to determine who will obtain a seat in parliament (Müller, 2005: 410), electoral competition is party-centered. Survey research shows that Austrian voters think of elections in terms of parties rather than individual candidates (Müller, 2005: 410). This make sense because voters decide between parties rather than individual candidates at the ballot box. Parties fill their legislative seats with candidates in the order they appear on the party's list.

To win office in Austria, a candidates' best strategy is to curry favor with the party's leadership in order to obtain a high place on the party's list. Candidates higher on the party's list are more likely to win office than those lower on the list. Therefore, to maximize their chances of winning office, candidates competing in party-centered PR systems, such as Austria, work to appease the party leadership in order to win a high place on the party's list. To achieve this goal, legislators typically tow the party line – even if it conflicts with the interests of voters' in their geographically defined district. In other words, legislators prioritize the demands of the party leadership over those of voters in their districts.

In contrast, politicians competing in candidate-centered systems work to appeal to voters in their own electoral district. To maximize their chances of (re)election, politicians competing in candidate-centered elections may prioritize the demands of their constituents over those of party leaders. In candidate-centered systems, politicians can go against the party line because they have direct access to the ballot, and citizens cast votes for individual candidates rather than political parties. But in countries like Austria, where parties fill their legislative seats with candidates in the order they appear on the party's list, candidates tend to tow the party line. Given this, I expect Austrian legislators will support their party's position on the farm-gate tax break.

Implications for Subsidies

Two political parties advocated the farm-gate tax break: the SPÖ (Social Democrats) and ÖVP (Austrian Peoples Party/Conservatives). These parties made up the government coalition that proposed the tax break.[53]

---

[53] The Socialist Party of Austria (SPÖ) won 43.3 percent of the votes in Styria (Steiermark) in the October 7, 1990 General Election. The next closest party was

Under the responsibility of the Finance Minister, the Cabinet agreed to the tax break and the government subsequently presented the proposal to parliament for approval. Speaking in parliament, the Finance Minister said,

"When I am thinking of the beverages consumption tax I have to say that we found a compromise on a topic that has been debated for decades, a compromise that gives equal opportunities to retailers and tourism."[54]

Finance Ministers are typically less supportive of tax breaks than other government ministers because they reduce government revenues.[55] Yet, in this case, the Finance Minister supported the tax break and cited as a reason for his support the opportunities it would provide to two geographically diffuse groups: farm-gate wine retailers and the tourism industry.[56] Farm-gate wine merchants are spread across all of Austria's wine regions. Similarly, the tourism industry employs people across virtually the entire country. The tourism industry is, in fact, one of the most geographically diffuse industries in Austria. The farm-gate tax break constituted a potential boon for the geographically diffuse tourism industry. It was widely believed that consumers would travel to farm-gate wine merchants in order to take advantage of the cheaper wine made possible by the tax break.

For some legislators, their party's position on the tax break corresponded with their constituents' economic interests. Legislators from wine regions who were members of a government party, for example, faced congruent incentives to support the tax break. Both their party and their constituents wanted the tax break and as a result, these legislators found themselves in a win-win situation. Their winemaking (and wine-consuming) constituents would benefit from the tax break, and they would earn credit with the party leadership by supporting the tax

---

Austrian People's Party (ÖVP) with 33.2 percent of the vote. In the following election, support for the Socialist Party fell. The SPO received only 36.6 percent of the vote. The Austrian People's Party received 27.5 percent of the vote. The party with the largest vote gain was the Freedom Party of Austria (FPÖ). Its vote share increased to 23.4 percent up from 16.8 percent in 1990. Despite this, the Socialist Party of Austria (SPÖ) remained the largest party at the national level. In 1994, the socialists held sixty-five seats in the *Nationalrat*. This represented a seat lost from 1990 when then held eighty seats.

[54] Lacina, pp. 8511–8514, in Stenographic Protocol of the 77th session of the National Council of the Republic of Austria, 9–10.06.1992, Vienna (Org. Title: Stenographisches Protokoll der 77. Sitzung des Nationalrats der Republik Österreich, Wien).

[55] Particularly as compared to those ministries entrusted to promote industrial or regional development (Blauberger 2009).

[56] Stenographic Protocol of the 73rd session of the National Council of the Republic of Austria, 24.06.1992, p. 7934, Vienna (Org. Title: Stenographisches Protokoll der 73. Sitzung des Nationalrats der Republik Österreich, 24.06.1992, 7934, Wien).

break. Erhard Meier was one such legislator. Meier was a member of the Social Democrats (SPÖ) party – one of the government parties that proposed the tax break. Meier represented the state of Styria in Austria's second chamber, the *Bundesrat* (Federal Assembly).[57] Styria is the second largest wine producing state in Austria. Nearly 3,300 (3,290.83) hectares were devoted to vineyards in Styria in 1999 (Statistics Austria 2000). For comparison, the next largest wine-producing area (Vienna) had only 678.30 hectares of vineyard (Statistics Austria (2000).[58] Given the importance of wine production in his home state and his party's support of the subsidy, it is no surprise that in December 1991, Meier spoke in support of the subsidy in parliament.[59]

By speaking in support of the subsidy, Meier promoted the economic interests of all farm-gate wine merchants – not just those in his own state of Styria. Given this, some part of Meier's efforts would go unrewarded in a plurality system like France because many of the beneficiaries of the subsidy could not vote for him. However, in Austria, Meier's work was not in vain. Although farm-gate wine merchants outside of Styria could not vote for Meier, they could vote for his party. And Austria's three-tier PR electoral system ensures that no votes are wasted. All votes count towards the party's share of legislative seats. By speaking in support of the tax break, Meier promoted his party's policy and potentially improved their future electoral success. By doing so, he also curried favor with the party leadership, which is an astute strategy in a closed-list PR system like Austria where parties control access to the ballot and the identity of people who fill seats in the legislature.

Unlike Meier, some legislators faced a situation in which their party's position on the tax break conflicted with the economic interests of their constituents. Erich Schreiner (Freedom Party of Austria/Freiheitliche Partei Österreichs/FPÖ), for example, represented the country's largest wine producing region (Lower Austria) with over 44,567.53 hectares of vineyards in 1999.[60] Many of Schreiner's constituents would benefit from the farm-gate tax break. Yet, Schreiner's party, the FPÖ, opposed the tax

---

[57] Prior to joining the legislature, Meier served as mayor of Bad Aussee (1975–1990), a town in the Austrian state of Styria. He also served as District Party Deputy Chairman of the SPÖ Liezen, a municipality in the Austrian federal state of Styria and as a member of the Land party committee of the SPÖ Styria since 1987.

[58] While the number of hectares under cultivation has changed over time, the relative ranking of Austrian states has not (Statistics Austria 2000). The 1999 data are the closest available data to the tax break.

[59] Ultimately, the 1992 Wine Tax Act included an exemption of for farm-gate wine sales.

[60] While the number of hectares under cultivation has changed over time, the ranking of Austria's wine producing regions has not (Statistics Austria 2000).

break. Unsurprisingly, given Austria's electoral institutions, Schreiner chose to support his party's position rather than the economic interests of his geographically defined constituents. In a parliamentary debate over the Wine Tax Act in the National Council on June 1992, Schreiner towed his party's line and argued against the subsidy saying,

someone in a wine growing region can go to a farm gate merchant and buy a cheaper product because it is less taxed, while someone in Salzburg has to pay a higher price.[61]

The Austrian state of Salzburg is located in the north of the country, close to the border with the German state of Bavaria. Salzburg is part of Austria's beer belt, which runs on the northern border from Waldviertel (the Wood Quarter) in Lower Austria through all of Upper Austria, Salzburg, Tyrol to Vorarlberg.[62] Although historically there were vineyards in Salzburg, virtually no wine is produced in the region today because of its current climate. While consumers in Austria's wine growing regions, including Eastern Austria, Burgenland, and Styria would benefit from the subsidy thanks to their geographic proximity to farm-gate wine merchants, consumers in Salzburg would benefit relatively less because they live farther away from farm-gate wine merchants.

Although Schreiner argued for the interests of citizens of Salzburg, he was not elected to represent Salzburg. Instead, he was elected to represent voters in Lower Austria and specifically the Waldviertel (Wood Quarter) where Austria's most famous wine region, the Wachau, is located. By opposing the tax break, Schreiner promoted the position of his party at the expense of his winemaking (and wine-drinking) constituents. Schreiner's decision is eminently rational in a closed-list, party-centered system like Austria. But why identify Salzburg specifically in his parliamentary speech?

Schreiner's reference to Salzburg was no coincidence but instead was part of the FPÖ's electoral strategy. The FPÖ, stood a greater chance of directly winning an additional legislative seat in Salzburg than in Lower Austria. In both Lower Austria and Salzburg, the FPÖ was the third largest party by vote share. However, the FPÖ was much closer to its next largest competitor, the Austrian People's Party (ÖVP) in Salzburg. In the 1990 parliamentary elections, less than 12 (11.6) percentage points separated FPÖ from ÖVP in Salzburg.[63] In contrast, nearly 27 points

---

[61] Schreiner, pp. 8555–8557, in Stenographic Protocol of the 77th session of the National Council of the Republic of Austria, 9–10.06.1992, pp. 8554–8558, Vienna (Org. Title: Stenographisches Protokoll der 77. Sitzung des Nationalrats der Republik Österreich, 9–10.06.1992, pp. 8554–8558, Wien).
[62] Email correspondence with Professor Wolfgang Müller February 26, 2016.
[63] The FPÖ won 20.5 percent of the votes while ÖVP won 32.1 percent of the vote in Salzburg.

separated the FPÖ from ÖVP in Lower Austria. The FPÖ therefore stood a better chance of gaining an additional seat in Salzburg relative to Lower Austria.

The FPÖ leadership calculated that supporting the tax break would do little to shrink the 27-point gap between it and its next largest competitor in Lower Austria: ÖVP, who supported the tax break. FPÖ was unlikely to make any gains by supporting the subsidy relative to ÖVP given that ÖVP helped to draft the proposed legislation.

By opposing the tax break, the FPÖ sought to appeal to a large group of voters (i.e. consumers) in a state where they were relatively competitive (i.e. Salzburg). In other words, the FPÖ opposed the farm-gate tax exemption for strategic electoral reasons. By opposing the subsidy, the FPÖ stood to gain votes in in Salzburg where few voters would directly benefit from the subsidy.

Schreiner's objection to the tax break illustrates the contrasting significance of geography in different electoral systems. In plurality systems, politicians are willing – even eager – to confer geographically specific benefits, such as the subsidies to wine producers in Charente and Charente-Maritime. The French program made subsidies available to only a small number of winemakers in the demarcated "Cognac" area of Charentes. The French government estimated that just 1,000 hectares were eligible for the subsidy. Despite the very small number of potential beneficiaries, no opposition to the subsidy is recorded in the legislative record. No French legislators spoke out against this geographically targeted subsidy, even though it would confer benefits in a geographically unequal manner.

In contrast, Austrian legislators objected to the farm-gate subsidy precisely because of its geographic inequalities. The fact that not all Austrians could benefit equally from the subsidy was, for some legislators and political parties, a compelling reason to oppose it. Legislator Schreiner said, for example, that the geographic inequality generated by the tax break was, "not justified."[64] Schreiner opposed the subsidy despite the fact that a large number of farm-gate wine merchants lived and worker in the electoral district he represented in parliament.[65]

For French legislators, taking such a position would be unthinkable. French legislators work on behalf of their constituents. In fact, many French MPs believe it is their responsibility to represent their

---

[64] Schreiner, pp. 8555–8557, in Stenographic Protocol of the 77th session of the National Council of the Republic of Austria, 9-10.06.1992, pp. 8554–8558, Vienna (Org. Title: Stenographisches Protokoll der 77. Sitzung des Nationalrats der Republik Österreich, 9–10.06.1992, pp. 8554–8558, Wien).

[65] In 1990, he served as Regional Party Vice-Chairman of the FPÖ of Lower Austria.

constituents above all else. This explains why legislators elected from the Cognac region lobbied for subsidies for Cognac producers. Importantly, French legislators from other regions did not oppose the subsidy program even though their constituents could not benefit from it. Legislators from other regions did not query the fact that all of the benefits were going only to Cognac producers. Nor did they use the opportunity to agitate for general subsidies for all winemakers or for targeted subsidies for winemakers in their own district. Instead, French legislators from other regions gave their implicit consent to geographically targeted subsidies.[66]

Electoral institutions explain the difference in legislators' behavior. Different electoral institutions generate dissimilar incentives for leaders to cater to more or less concentrated groups. Austria's electoral institutions, which include closed-list PR and multimember districts, give parties incentives to cater to geographically diffuse groups, such as farm-gate wine merchants. Furthermore, in a party-centered system like Austria, legislators' best reelection strategy is to support their party's position – regardless of the economic interests of voters in their geographically-defined electoral district. Schreiner, for example, supported the position of his party rather than the economic interests of his winemaking (and wine-drinking) constituents. By opposing the subsidy, Schreiner increased his chances of obtaining a good position on the party-list. In Lower Austria, his party won four legislative seats in the 1990 elections. Schreiner could nearly ensure his reelection if he was at or very near the top of the party list. By actively working to promote the party's position on the tax break, he curried favor with the party leadership and maximize his chances of reelection. The party's position was chosen in an attempt to maximize their chances of directly gaining additional legislative seats. This example suggests that parties competing in multiparty PR systems may target their appeals to districts where they have the best chance of upsetting the rank ordering of parties.

The Policy Outcome

The Wine Tax Act came into force July 31, 1992 via the ordinary legislative procedure.[67] It was voted on after the 3rd reading of the act.

---

[66] Perhaps they did so as part of an iterated logroll knowing that others would "return the favor" when they faced demands for subsidies from their own constituents and would need to lobby for geographically targeted subsidies to appease them.

[67] Federal Law Gazette of the Republic of Austria No. 154, 30.06.1992, pp. 1759–1763, Vienna (Org. Title: Bundesgesetzblatt für die Republik Österreich Nr. 154, 30.06.1992, 1759–1763, Wien).

The last reading proceeded without debate immediately following the 2nd reading in the National Council on 10th July 1992[68] and in the Federal Assembly on July 15, 1992.[69] The tax exemption for farm-gate wine sales, included in the Wine Tax Act of 1992, passed by at least a two-thirds majority of votes in the National Council (which was needed since the Tax Act also included amendments to the constitution). It passed by unanimous vote in the Federal Assembly. Unfortunately, individual parliamentarians' votes on the legislation were not recorded.

Although the tax break was implemented prior to Austria's EU membership, the Commission ultimately ruled that it did not conform to EU rules.[70] The Commission's logic was that wine and other fermented beverages sold directly to the consumer at the vineyard compete with wine and other fermented beverages offered to the same consumer by retail traders. By exempting from charges one group of products while charging the others, conditions of competition were distorted. The Commission ruled that Austria could not maintain this subsidy after December 31,1998.

CONCLUSION

In this chapter, I examine two subsidy programs – one for winemakers in France and another for winemakers in Austria. These two cases illustrate the political mechanisms that link electoral institutions and economic geography to policy. Policy outcomes are shaped by incumbents' optimal reelection strategy, which depends on a country's electoral institutions and economic geography. In plurality systems, incumbents' best reelection tactic is to target economic benefits, like subsidies, to producers their own electoral district because doing so maximizes their chances of reelection. Precisely for this reason, legislators from the French region of Cognac spoke in parliament to demand subsidies exclusively for winemakers in their districts. In contrast, legislators in Austria did not always represent their constituents' economic interests in parliament. One legislator, for example, spoke against a subsidy that would directly benefit winemakers in his district. He did so because his reelection chances

---

[68] Stenographic Protocol of the 77th session of the National Council of the Republic of Austria, 9–10.06.1992, pp. 8554–8558, Vienna (Org. Title: Stenographisches Protokoll der 77. Sitzung des Nationalrats der Republik Österreich, 9–10.06.1992, pp. 8554–8558, Wien).

[69] Stenographic Protocol of the 557th Session of the Federal Assembly of the Republic of Austria, 15.07.1992, pp. 26637–26640, Vienna (Org. Title: Stenographisches Protokoll der 557. Sitzung des Bundesrats der Republik Österreich, 15.07.1992, pp. 26637–26640, Wien).

[70] Beverage tax exemption farm gate; case number C57/96 http://eur-lex.europa.eu/legal-content/EN/TXT/?uri=CELEX:31999D0779.

depend on his party's electoral fortunes and his support among the party's leaders because of Austria's de facto closed party lists.

As these two cases illustrate, electoral institutions and economic geography together shape incumbents' (re)-election strategies and subsequently economic policy. Incumbents in plurality systems, like France, lobby for subsidies for producers concentrated in their own electoral districts to maximize their reelection prospects. In contrast, government parties elected via proportionality and closed party lists, as in Austria, fund subsidies for geographically diffuse producers to maximize their chances of staying in government. Although the findings from these two cases are consistent with those from the large-N studies, several novel points emerge that merit discussion.

First, not all government-funded subsidies are equally "narrow" and consequently researchers must accurately identify a policy's beneficiaries to fully understand the politics behind it. Subsidies are generally considered to be "narrowly beneficial" policies. Yet, the two programs I investigate here demonstrate the potential variation in subsidies' beneficiaries. Both subsidy programs aided a small subset of the total population. However, the scope of the beneficiaries varied meaningfully between the two programs. The French subsidy benefited select producers on just 1,000 hectares of land. In contrast, the Austrian subsidy helped all farm-gate wine merchants across the entire country. In this way, the Austrian subsidy was more broadly beneficial than the French subsidy.

Accurately classifying policies as being either "broadly beneficial" or "narrowly targeted" has long been a stumbling block for empirical research on the form of government policy (e.g. Cox and McCubbins 2001, Persson and Tabellini 2003, Golden and Min 2013). In this chapter, I offer a potential way forward. Rather than using policy categories, such as subsidies, as imprecise proxies for the breadth of beneficiaries, scholars should instead work to identify the actual beneficiaries of government policies (e.g. farm-gate wine merchants) and their geographic distribution. Such a task is difficult but not impossible, particularly given recent advances in data transparency.[71]

Second, economic policy may depend, in part, on the partisanship of a country's government. Socialist governments enacted both of the programs investigated here. However, the large-N quantitative studies described in other chapters generally show no systematic relationship between partisanship and subsidy spending. Left governments do not spend any more of their budgets on subsidies than right governments, holding all else equal. Perhaps left and right governments spend similar

---

[71] See, for example, Jusko (2017).

amounts of money on subsidies but assist different industries. Right-leaning parties may subsidize capital-intensive industries while left-leaning parties may assist labor-intensive industries (Verdier 1995). Because wine production is relatively labor-intensive, socialists may be more inclined to subsidize winemakers than other political parties.[72] If different parties subsidize different industries, then it is important to account for government partisanship when comparing subsidies between industries.[73]

Third, electoral competitiveness may influence the geographic distribution of subsidies within countries. In France, legislators who won office with smaller margins were more likely to lobby for subsidies than legislators who won with big margins. This may be because legislators can point to their legislative efforts at the next election and say, "See, I worked hard to get you subsidies. You should vote for me." In contrast, legislators who command safe majorities feel less compelled to lobby for subsidies. Incumbent politicians who won a large share of a district's votes in the previous election do not need to chase after additional votes because each new vote only contributes to their already large margin. As a result, subsidies may go disproportionality to districts with more competitive elections (i.e. swing seats) in plurality systems.

Although the French case suggests that district-level electoral competitiveness may influence the geographic distribution of subsidies in a plurality country, it is unclear what impact, if any, competitiveness may have on subsidies in PR countries. Existing research on competitiveness is generally restricted to plurality systems and, in fact, competitiveness most often refers to elections in single-member districts.[74] But in Austria – a proportional system with multi-member districts – competitiveness calculations may have led a party to oppose subsidies. An opposition party opposed the subsidy proposal to appeal to voters in Salzburg where they stood a greater chance of winning an additional legislative seat than in other districts. To win the votes needed to win an additional seat, the party spoke out against the subsidy because few, if any, people in Salzburg would benefit from it. Intriguingly, the Austrian example suggests district-level electoral competitiveness may

---

[72] However, in France, governments of various ideological persuasions have assisted the wine industry.
[73] Single industry studies hold constant industry-specific characteristics such as labor-intensiveness and as a result, government ideology is not required as a control variable.
[74] In these studies, competitiveness most often refers to district-level competition rather than country-level measures of competitiveness. However, see, for example, Monroe and Rose (2002). But even at a district level, it is not always clear what "competitiveness" means in PR systems.

have previously overlooked effects on economic policy in PR systems. In Chapter 7, I investigate this possibility by examining the impact of district-level electoral competitiveness on the distribution of subsidies between electoral districts in an archetypal PR country: Norway. First, however, I examine the variation in subsidies between countries with proportional electoral systems. In the next chapter, I also examine the variation in subsidies between sectors within a PR country.

6

Why Institutional Differences among Proportional Representation Systems Matter

Governments in different countries spend wide-ranging amounts of money on subsidies. Even among countries with similar electoral rules, governments allocate varied amounts of money to subsidies for business. I argue that governments elected via similar electoral systems spend different amounts of money on subsidies because of economic geography. However, economic geography alone cannot account for the entire variation in subsidies between countries with similar electoral rules. Among countries with proportional electoral rules, some governments spend relatively more on subsidies, holding constant the geographic dispersion of recipients. What explains such variation?

One possible explanation is the institutional diversity that exists among countries classified as having PR systems. The blunt distinction between plurality and proportional systems masks significant institutional variation. Among countries classified as having proportional systems, seven distinct electoral formulas are used to allocate seats to parties. Proportional systems also differ in their district magnitude, the use of higher electoral tiers, the use of electoral thresholds, and the type of party list employed (Clark et al., 2013: 543, Gallagher, Laver, and Mair, 2006: 354).[1]

This institutional diversity raises questions about the usefulness of the blunt distinction between proportional and plurality systems. Indeed, some scholars warn that only nonspecialists of electoral systems can persuade themselves that all electoral systems can be characterized by

---

[1] Although most PR systems use some type of list, not all PR systems do so. In single transferable vote systems, candidates' names appear on the ballot, often in alphabetical order, and voters rank at least one candidate in order of their own preferences. Candidates that surpass a specified quota of first-preference votes are immediately elected. In successive counts, votes from eliminated candidates and surplus votes from elected candidates are reallocated to the remaining candidates until all the seats are filled (Clark et al., 2013: 578). Such a system is used, for example, in Ireland to elect members to the lower chamber of the *Oireachtas*.

a single dichotomous indicator (Taagepera and Qvortrup 2011). They call the plurality/proportional distinction a "procrustean bed" (Taagepera and Qvortrup, 2011: 255) and warn researchers to, "forget about such coarse dichotomy" (Taagepera and Qvortrup, 2011: 253). The blunt dichotomy between proportional and plurality electoral systems may obscure important institutional variations that matter for countries' economic policies.

The institutional diversity among countries characterized as having proportional systems may explain why governments elected via PR spend varied amount on subsidies, even accounting for economic geography. Governments may spend more on subsidies in some PR countries than others because of institutions that exist only in certain PR systems. In this chapter, I investigate two institutions that vary among PR systems: list type and district magnitude. These institutions may generate diverse incentives for governments elected via proportionality to supply subsidies and other particularistic economic policies.

List type refers to how candidates fill a district's multiple seats. In open-list systems, candidates fill a party's seats in the order in which they are ranked by voters. In closed-list systems, candidates are chosen by party leaders to occupy a party's legislative seats.

List type shapes politicians' (re)election tactics and subsequently their policy decisions. In open-list systems, politicians win office by currying favor with voters in their own district. Voters can express a preference for a particular candidate on an open party list. Candidates ranked higher by voters have a better chance of winning a seat in the legislature. To build personal support among voters, politicians may promise subsidies to groups concentrated in their own district (or bailiwick). Once elected, legislators will work to fund the promised subsidies to maximize their chances of reelection. Given this, I hypothesize that subsidies for geographically concentrated beneficiaries will be more generous in open-list systems than closed-list systems, all else equal.

In contrast, candidates in closed-list systems have little incentive to champion subsidies for groups concentrated in their electoral district. Voters cannot express preferences for individual candidates in closed-list systems. Candidates' names are often not even included on the ballot in closed-list systems. Instead, voters select a party. Party leaders then decide which candidates will fill the seats allocated to the party. Because party leaders choose who fill will a party's legislative seats, candidates' seek to curry favor with party leaders. Party leaders often champion programs for geographically diffuse groups in order to maximize the party's electoral success and candidates will support these programs to please party leaders. As a result, I hypothesize that spending on subsidies for geographically diffuse groups will be higher in closed-list PR systems, as compared to open-list PR systems. In open-list systems, some funds will diverted from diffuse groups to geographically concentrated groups by candidates seeking to build a personal support base.

District magnitude refers to the number of candidates elected in a district.[2] In proportional systems, multiple candidates are elected from each district. However, the number of candidate elected from each district varies within and between PR countries.[3] In some countries, as many as 150 legislators are elected from a single district (e.g. Slovakia). In others, as few as two legislators are elected from each district (e.g. Chile).

District magnitude may interact with a country's list system to shape politicians' incentives to supply certain types of economic policies. In districts with more seats, politicians compete against more politicians from their own party because parties tend to run more candidates in districts with more seats. In theory, a party might run as many candidates as there are seats. In a district with five seats, for example, a party might run five candidates in the hope of winning all five. Although practical constraints may render this strategy infeasible (particularly for smaller parties), the number of candidates running for office from a given party in a district generally increases with district magnitude. When district magnitude is high, candidates compete against relatively more co-partisans for voters' support, and consequently they must work harder to distinguish themselves from other candidates. One way to distinguish themselves from other co-partisans is to provide subsidies. This logic suggests that increases in mean district magnitude will correlate with greater government spending on subsidies to concentrated groups in open-list PR systems.

I test these hypotheses using cross-nationally comparable data on government-funded subsidies in countries with proportional electoral systems. I supplement the cross-national tests with a single-country study. A single-country case study holds constant cultural and institutional features, such as list type, and consequently helps to isolate the effects of economic geography on policy. Economic geography varies between sectors within a country. In other words, some sectors in a country will be more geographically concentrated than others. By exploiting the within-country variation in economic geography, I can test the effects of economic geography on policy outcomes while holding constant other country-specific features. Using this research design, I find

---

[2] District magnitude is distinct from the geographical size of a district.
[3] While plurality systems typically elect a single legislator from each district, district magnitude varies among proportional systems.

that governments elected via proportional rules and closed party lists spend more on subsidies for geographically diffuse sectors, as compared to geographically concentrated sectors. The single-country results confirm the findings from the multiple-country regressions.

VARIATION IN LIST TYPE

In most PR systems, party lists determine how candidates are chosen to fill a party's legislative seats.[4] Two distinct types of lists exist: open and closed. Closed lists give party leaders exclusive control over which candidates will fill the party's legislative seats. Candidates are seated strictly according to the order in which the party has ranked them on their list. Candidates closer to the top of the party list are more likely to get a seat in parliament. In this way, party leaders, rather than voters, decide which candidates will represent citizens in the legislature. Voters have no direct ability to affect which of the party's candidates actually represent them (Cox and McCubbins 2001). At the ballot box, voters cannot express a preference for any individual candidate (Lijphart, 1999: 147). In fact, ballot papers in closed-list systems frequently do not even list the names of individual candidates. Instead, voters choose a party knowing that the party leaders will fill the legislative seats won by the party (Cox and McCubbins 2001). Examples of closed-list PR systems include Spain and Israel.

In open-list systems, parties cannot fully control the order in which candidates receive seats. Instead, voters have a say over which candidates will represent them in parliament. Legislative seats are allocated to parties based on the sum of the votes for all the candidates of a given party. If a party wins four legislative seats, the candidates seated are the four on the party's list that received the most individual preferences votes (Cox and McCubbins 2001). In contrast, in a closed-list system, they would be the four candidates that the party leaders placed in the top four positions on its list (Golden and Picci 2008). Finland is an example of a fully open-list PR system. Many other countries use partially open or "flexible" lists where voters can express preferences for individual candidates but the list order presented by the parties often prevails because voters decide not to express a preference between candidates or because of threshold

---

[4] Single transferable vote systems do not use a party list. Such a system is used, for example, in Ireland to elect members to the lower chamber of the Oireachtas. However, in my sample, I code Ireland as being an open-list system because although parties control access to the ballot, voters choose which candidates will represent them in the legislature and elections are candidate-centered. All reported results are robust to the exclusion of Ireland from the sample.

requirements (Lijphart 1999). In Sweden, for example, a candidate needs to receive 5 percent of the party's votes in order for "individual preference votes" to overrule the ordering on the party list.

THE POLICY EFFECTS OF LIST TYPE

In both open and closed-list PR systems, parties control access to the ballot. In closed-list PR systems, parties also fully control which candidates will fill the legislative seats the party wins. In open-list systems, voters have a say over which candidates will represent them in parliament. This seemingly small institutional difference has important consequences for electoral behavior and subsequently policy.

Closed Lists

The type of party list influences candidates' election strategies and the nature of electoral competition. Closed lists generate party-centered competition, which encourages voters to emphasize their party preference over that for specific candidates. Because voters cannot express a preference for individual candidates at the ballot box in closed-list systems, they must prioritize their party preferences. Voters decide which party to support based on the parties' platforms and election promises – rather than any individual candidate's personal characteristics.

In addition to having an effect on electoral competition, list type also has an effect on legislative behavior. Closed lists typically engender high levels of party discipline (Depauw and Martin 2008). Party discipline refers to the control that party leaders have over their members of the legislature. In closed-list systems, party leaders can exert control over their members because legislators' best chance of winning reelection is to be at or near the top of the party list. Party leaders can effectively promise candidates a legislative seat by placing them at the very top of the party's list. To earn a position at the top of the list, legislators work hard to appease the party's leaders by promoting the interests of the party. In closed-list systems, fighting for the interests of their geographically defined constituents (i.e. voters in their own electoral district) does little to improve legislators' reelection chances – especially when the interests of their constituents run counter to the interests of the party. Legislators who work against the party's interests may be moved down the party list, which reduces their chances of winning reelection. Closed-list systems consequently engender high levels of party discipline.

Closed-list systems also align the incentives of individual legislators and party leaders. Party leaders seek to maximize the number of legislative seats the party holds. Individual legislators also work to promote their party, which in turn maximizes the party's legislative seats. This dynamic reinforces the party-centered nature of electoral competition in closed-list systems.

Closed lists generate party-centered electoral competition, high levels of party discipline, and align the incentives of legislators and party leaders. As a result, constituency-focused earmarks like those frequently seen in plurality countries are improbable in closed-list systems. A city that wants money for a new museum, for example, might lobby their local representative(s) for government funds. But legislators elected via closed-lists have few incentives to respond to the demands of their geographically defined constituents – particularly if the legislator's own party has no interest in such a program. The party may instead want to spend government funds on other projects – specifically those most likely to maximize the party's legislative seat share. In this way, geographically concentrated interest groups often find themselves without a champion in closed-list PR systems – particularly at the national level. As a result, geographically concentrated producers tend not to win generous government subsidies in closed-list systems.

Instead, geographically diffuse groups are relatively more likely to win subsidies in closed-list systems. Subsidizing a geographically diffuse sector helps people across the country. For example, the Norwegian construction industry employs people across the entire country and subsidizing this industry consequently helps citizens in all regions. Employees in the construction industry benefit directly from subsidies via increased wages and more secure employment. Owners of capital invested in the industry benefit from above market rates of return and greater demand. Subsidies to the construction industry also help related sectors, such as real estate and retail. In this way, subsidies to one industry can indirectly benefit connected industries and services (Barber 2014).

Subsidies to a geographically diffuse industry indirectly benefit many more people than subsidies to a concentrated industry. In effect, there is a "dispersion bonus" from subsidizing geographically diffuse industries. The political profit from this dispersion bonus is greater for parties competing in PR systems, as compared to plurality systems. Parties in PR systems seek to maximize the number of votes they win in order to maximize the number of seats they hold in the legislature. In this institutional setting, subsidies to diffuse groups are a politically expedient electoral tool.

Even subsidizing a relatively small, yet diffuse, industry could be electorally beneficial in a PR system. If, for example, an industry employs just 2 percent of the population, a party could potentially increase its vote share by 2 percent by subsidizing that industry. Depending on the magnitude of the "dispersion bonus" the electoral gains may be even larger. But even an increase of just 2 percent could be electorally valuable for parties competing in PR systems. In Sweden, for example, a 2 percent increase in a party's vote share could translate into as many as seven additional legislative seats.

In all PR system, the political bonus from subsidizing geographically diffuse groups is electorally valuable for parties. Yet, only parties in closed-list systems have the necessary control over their legislators to ensure that funds go to the diffuse groups that maximize the party's vote share. Party discipline in closed-list PR ensures that subsidies flow to geographically diffuse groups rather than the geographically concentrated supporters of individually powerful legislators (Golden and Picci 2008). In contrast, parties in open-list systems have greater difficulty disciplining their legislators' vote-seeking behavior, as described in the following section.

Open Lists

Open lists generate candidate-centered electoral competition. Candidate-centered competition encourages the voter to see the basic unit of representation as the candidate rather than the party (Shugart, 1999: 70). Candidate-centered competition emerges in open-list systems because voters indicate not only their preferred party but can also designate their favorite candidate within that party (Shugart, 1999: 70). The party's legislative seats are filled by those candidates who win the most votes. In this way, candidates from the same party compete against one another for votes and ultimately seats. Legislators cannot guarantee their own reelection just by working to ensure the party's popularity. While a popular party may win many seats in parliament, voters ultimately decide who fills a party's seats in open-list systems.

To maximize their chances of reelection, incumbents must do something to distinguish themselves from other candidates on a party's list. One thing they can do is develop a personal vote. A personal vote is that part of a legislator's vote that is based on his or her individual characteristics or record rather than the party to which the candidate belongs (Carey and Shugart 1995). Developing a personal support base among voters helps candidates to differentiate themselves from others on the same party list. In this way, generating a personal vote maximizes candidates' chances of winning a legislative seat in open-list systems (Carey and Shugart 1995, Cox and McCubbins 2001).

Legislators can use particularistic economic policies to develop a personal vote (Lancaster 1986, Lancaster and Patterson 1990, Carey and Shugart 1995). Subsidies are one type of particularistic economic policy. Subsidies keep people employed, create new jobs and raise wages above market rates. By delivering subsidies, politicians can garner the support of voters who benefit directly, or even indirectly, from the program. In this way, subsidy spending is roughly analogous to legislative particularism, or "pork barrel" spending, when the beneficiaries of a subsidy are geographically concentrated. Subsidies to geographically concentrated groups represent the appropriation of government spending for localized projects secured solely or primarily to bring money to a representative's district. Bringing "pork" back to the district helps politicians cultivate a personal vote (Fenno 1978, Ferejohn 1974, Wilson 1986).

In open-list systems, individual legislators' best reelection strategy differs from parties' optimal electoral tactics. Individual legislators endeavor to develop a personal vote by targeting benefits to voters concentrated in their electoral district or bailiwick. In contrast, parties seek to maximize the number of legislative seats they control. At times, these two objectives will come into conflict. A legislator may want to subsidize producers in her district to cultivate a personal vote. Yet, the legislator's party may have no interest in subsidizing producers in that district. If, for example, the district is heavily populated by supporters of other parties, the legislator's party will have little interest in providing subsidies to producers in that district.[5]

This disconnect between legislators' incentives and parties' incentives engenders problems for parties in open-list systems. Party leaders have few tools with which to discipline their legislators in open-list systems. While parties can effectively "promise" a candidate a legislative seat in closed-list systems by placing them at or near the top of the party's list, the same is not true in open-list systems. The top spot on a party's list does not guarantee a seat in open-list systems because voters can upset a party's rank-ordering of candidates. As a result, parties have less ability to discipline their legislators' behavior in open-list systems.[6]

Because parties cannot wholly discipline their legislators' vote-seeking behavior in open-list systems, some subsidies will go to geographically concentrated groups. In open-list systems, legislators' best reelection strategy is to develop a personal vote by supplying particularistic

---

[5] See Chapter 7. See also Cox and McCubbins (1986), McGillivray (2004), Golden and Picci (2008).
[6] See Martin (2014), for other tools parties can use to discipline their legislators.

policies, like subsidies, to their geographically defined constituents. Parties cannot fully discipline such behavior and as a result some rents will flow to geographically concentrated groups in open-list systems. Evidence of this pattern is found in Italy during the period from 1953 to 1994 when open lists were used to elect national legislators (Golden and Picci 2008).[7] To cultivate their own personal support base, individually powerful legislators secured resources for their constituents in the form of infrastructure spending (Golden and Picci 2008). They did so even at the expense of the governing parties (Golden and Picci 2008). As the evidence from Italy illustrates, legislators in open-list systems have both the incentives and, at times, the opportunity to target benefits to their geographically defined constituents. Given this, I expect spending on subsidies to geographically concentrated groups to be higher in open-list systems, as compared to closed-list systems, all else equal.

CROSS-NATIONAL TESTS

I investigate the empirical relationship between list-type, economic geography, and subsidies in two different ways. First, I conduct a cross-national test using comparable data on central government spending on manufacturing subsidies for twelve countries with proportional electoral systems and varied list types. Second, I examine government spending on subsidies for different economic sectors in a single PR country with de facto closed lists. I describe these tests and the results in the following section and then examine the role of district magnitude.

Measuring List Systems

To capture the distinction between open and closed-list systems, I construct a variable that indicates when parties control both access to the ballot and the order in which candidates fill the party's legislative seats. When these two conditions are met, a country is characterized as having closed lists and the variable *Open List* equals zero. When parties control access to the ballot but not the order in which candidates receive seats, the country is classified as having open lists and the variable *Open List* equals one. These data come from Johnson and Wallack (2012) who build upon canonical insights from Carey and Shugart (1995).

---

[7] In Italy during this period, voters could decide to use as many as three (and in very large districts, four) preference votes for individual candidates on the party list of their choice (Chang and Golden 2007). Preference votes were restricted to one in the 1992 parliamentary elections.

The sample includes only PR countries because I am interested in the variation in governments' economic policies among proportional systems. While list type and district magnitude vary among PR systems, plurality systems have only one representative per district and often allow independent candidates and/or use primaries to select candidates. For this reason, plurality systems are excluded from the sample under investigation in this chapter. Although plurality systems are excluded, mixed member proportional (MMP) systems, like Germany, are included in the sample. In mixed systems, legislators in the same legislative chamber are elected via different institutions. To account for this, I estimate the institutional setting for the "average legislator" by calculating the weighted average of the variable *Open List*. Germany's value of *Open List*, for example, equals a country-level weighted average, which ranges from 0.95 to 0.99. All other countries in the sample take a value of zero or one. In Spain, for example, the variable *Open List* equals zero because party leaders determine which candidates will fill the party's seats (i.e. it is a closed-list system).

Although the coding scheme is straightforward, it is often difficult to determine the extent to which parties control candidates' order on the ballot in practice. In some cases, voters may disturb party lists with preference votes ("flexible lists"), but the extent to which they are actually affected in practice by preference votes varies according to thresholds and empirical circumstances. In Norway, for example, voters may, in theory, modify the order of candidates on the list. Voters are allowed to change the rank order of the candidates on the list as well as cross out candidates (Aardal, 2011: 8). However, the levels of coordination required to overturn the parties' rankings are so extreme that they effectively deter any attempts to do so. At least half of the voters have to make exactly the same alterations of the list for it to have any effect (Aardal, 2011: 8). As a result, Norway's system is effectively a closed-list system for all practical purposes (Aardal, 2011: 8).

The variable *Open List* captures de facto practice rather than de jure rules.[8] Flexible lists are coded as closed lists when, as in Norway, there is

---

[8] Johnson and Wallack (2012). The determination of ballot access is complicated. Oftentimes, where ballot access is legally controlled only by political parties, individual candidates may have de facto ballot access if they establish a new political party. Similarly, rules on the books regarding ballot access may imply that parties do not tightly control ballot access, but access is de facto prohibitive for other reasons, such as political dominance by a single-party or regime. To facilitate cross-country comparison of ballot access, Johnson and Wallack (2012) attempt to capture de facto practice that differs from de jure rules. Johnson and Wallack (2012) relied on the "Candidacy Requirements" listed on the Inter-Parliamentary Union's Parline database (www.ipu.org) to determine whether individual candidates faced stringent

144   *Why Institutional Differences among PR Systems Matter*

little or no actual change in list order based on electoral data and reports. Flexible lists are coded as open lists where preference votes actually influence which of a party's candidates are elected. Thirty-three percent of the sample observations use closed party lists, according to these criteria, while 59 percent use open lists. The remainder have a mixed system with a country-level weighted average that falls between zero and one, as in Germany.

Measuring Economic Geography

To estimate the geographic concentration of voters with a shared interest in manufacturing subsidies, I measure the concentration of manufacturing employment. Details about how this measure is constructed can be found in Chapter 4. The measure captures the degree of a sector's employment concentration relative to the geographic distribution of aggregate employment. The concentration variable ranges from zero to one with higher values indicating more geographic concentration. The construction of the geographic concentration variable is data intensive. Consequently, this variable can only be constructed for a relatively small sample of highly developed countries for which the necessary employment data are available. Although this limited sample raises potential questions about the external validity of the results, it allows for direct comparisons with previous studies of electoral institutions that use similar samples (e.g. Bawn and Rosenbluth 2006, Persson et al. 2007).

Measuring Subsidies

Subsidy spending equals the amount spent by the central government on grants and loans for the manufacturing sector as a percentage of total government expenditures (excluding interest payments).[9] This ratio

---

requirements for ballot access. Where available, Johnson and Wallack (2012) also inferred from electoral data or country reports the relative ease for individual candidates to appear on the ballot.

[9] These spending data come from the International Monetary Fund's *Government Financial Statistics*. These data include all fiscal outlays targeted to the manufacturing sector – including all subsidies, grants, and subsidized loans provided to the manufacturing sector to support manufacturing enterprises and/or development, expansion or improvement of manufacturing. Although conventional government accounts are generally not suitable for comparisons between countries and over time because they reflect the organizational structures of governments, these standardized data allow for meaningful cross-national comparisons. For additional information, see the *Government Finance Statistics Manual* (IMF 2001).

indicates the relative importance of subsidies among governments' myriad spending priorities.

Cross-National Model

I regress subsidy spending on employment concentration and an interaction term equal to the product of employment concentration and *Open List*, along with both constitutive terms. More precisely, a partial-adjustment regression is estimated by ordinary least squares (OLS) with the following form and robust standard errors:

$$\text{Subsidies}_{it} = \beta_0 + \beta_1 \text{Open List}_{it-1} + \beta_2 \text{Concentration}_{it-1} \\ + \beta_3 \text{Open List}_{it-1} {}^* \text{Concentration}_{it-1} + \beta X_{it-1} + \varepsilon_{it}$$

where ε_{it} is an error term. The coefficient on β_3 is expected to be positively signed; as the geographic concentration of manufacturing employment increases, I expect politicians in open-list systems to become relatively more responsive to demands for manufacturing subsidies.

X_{it-1} refers to a vector of control variables. All estimated models include at least three key control variables. The first control variable is a measure of trade openness. Since manufacturing subsidies assist domestic producers in competing with lower cost foreign imports, countries more open to trade may spend more on subsidies (Rickard 2012b). Trade openness may also systematically correlate with electoral institutions. Countries dependent on international trade tend to have electoral rules that minimize the influence of narrow, protectionist interest groups (Rogowski 1987). To reduce the potential for a spurious correlation, a variable measuring trade openness as the sum of imports plus exports divided by GDP is included as a control.

The second control variable included in all estimated models is *GDP per capita*. Electoral support from lower-income voters may be relatively cheaper to "buy" using subsidies (Lindbeck and Weibull 1987, Dixit and Londregan 1996). Manufacturing subsidies may, therefore, be higher in countries whose voters have relatively lower incomes on average as a result of strategic "vote-maximizing" spending by national governments.

The third control variable is country size, measured by the natural log of a country's land area in square kilometers. Large countries will tend to have bigger manufacturing industries, which may increase government spending on manufacturing subsidies. Country size may also relate systematically to electoral systems and the geographic concentration of sector employment (Blais and Massicotte 1997). Controlling for country size minimizes the possibility of finding a spurious correlation.

Despite the potential relationship among the three key control variables, standard tests show acceptable levels of multicolinearity.[10] Additional control variables are introduced one-at-a-time to further minimize multicolinearity. These additional control variables are:

Federalism, a dichotomous variable coded one for federal systems and zero otherwise. This is a potentially important control since the subsidy data refer only to central government expenditures. Data on general government spending, including that from local and regional governments, is often missing and when available it tends to be less reliable than central government spending data (Persson and Tabellini 2003). Furthermore, the precise definition of local and regional governments' outlays are often incomparable across countries and time periods (Persson and Tabellini 2003). Central government expenditures on subsidies may be lower in federal systems than nonfederal systems if some of the burden of subsidizing industries falls to regional and local governments. This would be particularly problematic if federal systems co-vary with electoral systems. *Federalism* is therefore introduced as a control variable.

Sector Employment equals the number of people employed in the manufacturing sector as a percentage of the total labor force. This is a potentially important control variable because the number of people employed in manufacturing may influence both the amount spent on manufacturing subsidies and the geographic distribution of manufacturing employees.

Left is a dichotomous variable coded one if the largest governmental party is left of center and zero otherwise. In general, governments' industrial policies tend to have only a minimal ideological component (Verdier 1995, McGillivray 2004, Thomas 2007, Rickard, 2012c). However, controlling for ideology is important because left governments tend to be associated with proportional electoral systems (Iversen and Soskice 2006). Failure to control for the ideological tendency of a government could result in mistakenly assigning explanatory power to electoral rules rather than ideology.

Production factors employed in geographically concentrated sectors may confront higher adjustment costs than factors in geographically diffuse sectors. Factors with higher adjustment costs invest more in lobbying (Hiscox 2002, Rickard 2009), which raises the possibility that asset specificity rather than *Concentration*, per se, explains groups' political influence. A measure of labor mobility is therefore introduced

---

[10] The variance inflation factor (VIF) is less than 4 for all variables included in the estimated models, as recommended by Huber et al. (1993).

as a control variable. This measure estimates the adjustment costs facing workers in the manufacturing sector by calculating the rate of labor movement between industries in the sector (Wacziarg and Wallack 2004). The rate of movement varies according to the costs to workers of voluntarily entering and exiting different industries. Higher rates of movement indicate lower adjustment costs. Including *Labor Mobility* in the estimated model does not change the key results and suggests that geographic concentration is, in fact, the source for explanation rather than adjustment costs.

Another plausible alternative explanation is the number of parties in government. The current study's argument maintains that electoral institutions affect politicians' incentives to cater to certain constituencies. Electoral systems may also influence subsidy spending via the number of parties in government. Single-party governments are most common in plurality systems, while PR systems are more likely to foster multiparty governments. Multiparty governments typically spend more than single-party governments because multiparty governments negotiate less efficient logrolls (Bawn and Rosenbluth 2006). This raises the possibility that multiparty governments will spend more on subsidies than single-party governments. If this is the case, the reported results may not be the consequence of the suggested electoral dynamics but rather the accountability and bargaining dynamics induced by multi- versus single-party governments. To test for this possibility, the number of parties in government is introduced as an additional control variable. Including the number of parties in government as a control variable in the estimated model does not change the key results.

Concentration (squared) tests for the possibility that maximum political influence occurs at some intermediate level of geographic concentration. The literature on interest group politics in plurality systems hypothesizes a positive coefficient for the un-squared concentration term and a negative coefficient for the squared term (Grier et al. 1994). However, the expectations are less clear for PR countries.

Cross-national Results

Governments elected via closed lists spend more on subsidies for geographically diffuse sectors than governments elected via open lists, as reported in Table 6.1. When geographic concentration is at its lowest observed value (i.e. zero), moving from a closed-list to an open-list PR system reduces subsidy spending by approximately one percentage point on average across all estimated models. The negative marginal effect of

Table 6.1 Effect of open-party lists on subsidy budget shares

	(1)	(2)	(3)	(4)	(5)	(6)	(7)	(8)
L.Open list	-0.274*	-0.949***	-0.963***	-1.049***	-1.152***	-1.371***	-0.915***	-1.027***
	(0.145)	(0.225)	(0.271)	(0.196)	(0.244)	(0.259)	(0.211)	(0.219)
L.Concentration	-23.83***	-36.12***	-36.32***	-39.00***	-39.27***	-46.01***	-25.03***	-53.78***
	(4.092)	(5.723)	(6.417)	(5.631)	(5.948)	(5.784)	(6.203)	(9.592)
L.Open list*L. Concentration		17.544***	17.643***	15.020***	19.700***	23.636***	17.456***	18.061***
		(6.057)	(6.359)	(5.778)	(5.974)	(7.666)	(5.518)	(5.799)
L.Trade	0.016***	0.017***	0.017***	0.015***	0.015***	0.016***	0.011*	0.018***
	(0.004)	(0.004)	(0.004)	(0.004)	(0.004)	(0.004)	(0.005)	(0.004)
L.GDP per capita (log)	-1.998***	-2.207***	-2.206***	-2.105***	-2.211***	-2.476***	-1.422***	-1.982***
	(0.197)	(0.206)	(0.206)	(0.204)	(0.201)	(0.226)	(0.342)	(0.262)
L.Area (log)	0.263***	0.299***	0.295***	0.292***	0.313***	0.347***	0.226**	0.322***
	(0.069)	(0.067)	(0.075)	(0.077)	(0.068)	(0.088)	(0.089)	(0.069)
Federalism			-0.021					
			(0.161)					
L.Employment				-1.413				
				(1.875)				
L.Left government					-0.272**			
					(0.130)			

(continued)

	(1)	(2)	(3)	(4)	(5)	(6)
L.Number of government parties				0.174**		
				(0.074)		
L.Labor Mobility					5.139	
					(6.279)	
L.Concentration^2						212.582**
						(101.720)
Constant	16.767***	18.786***	18.855***	19.226***	19.363***	22.095***	12.608***	16.595***
	(2.544)	(2.622)	(2.836)	(2.999)	(2.602)	(2.758)	(3.159)	(3.110)
Observations	162	162	162	141	162	141	144	162
R-squared	0.515	0.533	0.533	0.521	0.545	0.590	0.210	0.543

Robust standard errors appear in parentheses. All models include year fixed effects. Year coefficients are not reported due to space constraints. *** $p < 0.01$, ** $p < 0.05$, * $p < 0.1$.

open lists on subsidy spending is statistically significant for all values of geographic concentration less than 0.04. In other words, governments in closed-list systems spend more on subsidies than governments in open-list systems when employment is geographically diffuse. Seventy-seven percent of the sample observations fall in this range.

Spending on subsidies for geographically diffuse sectors is higher in closed-list systems as compared to open-list systems because parties in closed-list systems are better able to discipline their legislators' vote-seeking behavior. In open-list systems, where party discipline is lower, powerful individual legislators divert some money away from broadly beneficial programs to fund geographically targeted programs that benefit their own constituents. Legislators do this in an attempt to develop a personal vote, which is electorally valuable in open-list systems. In contrast, individual legislators in closed-list systems have neither the incentive nor the opportunity to funnel money to producers concentrated in their own district. Legislators and party leaders tend to eschew narrowly targeted programs in favor of programs that benefit people across the country in closed-list systems. As a result, spending on subsidies for geographically diffuse sectors is higher in closed-list systems than open-list systems.

As the beneficiaries of subsidies become more and more geographically concentrated, the negative marginal effect of open lists declines and eventually becomes positive, as illustrated in Figure 6.1. When

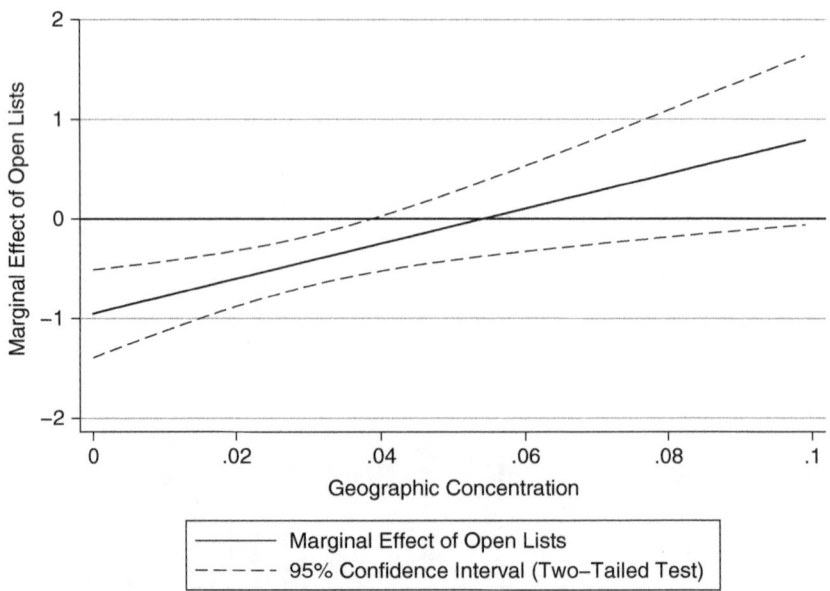

Figure 6.1 Marginal effect of open lists on subsidy budget shares in PR countries

geographic concentration is greater than 0.05,[11] moving from a closed-list to an open-list PR system increases subsidy spending. When geographic concentration is at its highest observed value, moving from a closed-list system to an open-list system increases subsidy spending by nearly one percentage point on average across all estimated models. Although the positive marginal effect does not reach the 95 percent level of statistical significance, it is consistent with evidence from previous studies of Italy and Brazil. Single-country studies have shown that electorally motivated policy targeting occurs in open-list PR. Voters in open-list PR countries, like Brazil, appear to support candidates offering particularistic economic policies, such as subsidies, over candidates promising broadly beneficial policies (Ames, 1995: 413).

As geographic concentration increases, governments in closed-list PR systems spend less on subsidies. This result is illustrated in Table 6.1 by the negative and statistically significant coefficient on *Concentration*. The coefficient on *Concentration* illustrates the effect of increased geographic concentration on subsidy spending in closed-list PR systems (i.e. when *Open List* equals zero). Across all estimated models, the coefficient is negative and significant indicating that government spending falls in closed-list PR systems as geographic concentration increases. More precisely, subsidy spending falls by 3.5 percentage points in closed-list systems when concentration increases from its minimum sample value to its maximum. The same increase in geographic concentration reduces subsidies by just 2 percentage points in open-list systems. Although geographic concentration is a political liability in all PR systems, it is relatively more detrimental to interest groups in closed-list PR systems where elections are party-centered and individual legislators have few incentives or opportunities to champion the interests of geographically concentrated groups.

I also estimate a two-stage least squares model to allay concerns about potential endogeneity. Indicators of the historical period during which a country's current electoral rules were adopted are used to instrument the variable *Open List* in the first stage of the model, following Persson and Tabellini (2003), Evans (2009) and others. The distribution of current electoral rules vary with the age of the rules (Persson and Tabellini 2003). Experience of other democracies and prevalent political and judicial doctrines shift systematically over time and these shifts may explain why the distribution of current electoral rules vary with the age of the rules. List type also varies with the age of electoral rules. Open lists are more common in younger electoral systems while closed lists are more frequent

---

[11] Which is the case for 16 percent of the sample observations.

in older systems. The age of electoral systems robustly predicts list type even though the type of list used generally changes more often than the electoral rule. Despite this, the age of a country's electoral system is a robust predictor of list type.

I exploit the temporal pattern in electoral systems by constructing three dummy variables that correspond to the periods: 1921–1950, 1951–1980, and post-1981. The dummy variables take a value of 1 if the current electoral system originated in the respective period, and 0 otherwise.[12] Countries' electoral systems and list type are robustly associated with the year in which countries' constitutions were adopted. However, the date at which a country adopts its constitution is unlikely to affect industrial policy or manufacturing subsidies. Indeed, the historical period during which a country's current electoral rules were adopted is not correlated with subsidy spending.

The results from the second-stage of the 2SLS model are reported in Table 6.2. The marginal effect of *Open List* on subsidies is negative and statistically significant when the beneficiaries are geographically diffuse. When *Concentration* equals zero, governments in closed-list systems spend more on subsidies than governments in open-list systems, all else equal. As *Concentration* increases, the negative marginal effect of *Open List* declines and eventually becomes positive. At high level of *Concentration*, the marginal effect of *Open List* is positive. In other words, governments in open-list systems spend more on subsidies than governments in closed-list systems when the beneficiaries are geographically concentrated.

In short, the key results are robust to an alternative model specification that relaxes the assumption that electoral systems are exogenous.[13] In fact, using the instruments in the first-stage model increases the estimated effects of key variables in the second-stage model. Compare, for example, the estimated coefficient on *Open List* in column 2. For the OLS results in Table 6.1, the estimated coefficient on *Open List* equals –0.95. For the 2SLS results reported in Table 6.2, the estimated coefficient equals –1.35. Instrumenting for a country's list type using the age of the electoral system increases by 42 percent the estimated reductive effect of *Open Lists* on subsidies to geographically diffuse groups. These results suggest that the OLS models likely report lower bound estimates.

---

[12] Persson and Tabellini (2003) demonstrate that these specific time periods best describe the pattern of electoral system adaptation.
[13] If anything, correcting for potential endogeneity appears to reduce the standard errors on the estimated marginal effect of *PR*.

Table 6.2 Second-stage results of the effect of open-party lists on subsidy budget shares

	(1)	(2)	(3)	(4)	(5)	(6)	(7)	(8)
L.Open List	−0.658***	−1.351***	−0.247	−1.399***	−1.630***	−1.794***	−0.813***	−1.101***
	(0.228)	(0.363)	(0.542)	(0.304)	(0.400)	(0.447)	(0.302)	(0.316)
L.Concentration	−32.47***	−46.92***	−31.93***	−48.22***	−51.53***	−57.47***	−28.71***	−59.23***
		(7.340)	(9.697)	(7.294)	(7.964)	(8.241)	(6.016)	(10.518)
L.Open List*L. Concentration		26.463***	9.142	24.977***	29.620***	33.917***	15.765**	20.302***
		(7.444)	(10.295)	(7.542)	(7.808)	(9.550)	(6.576)	(6.881)
L.Trade	0.012***	0.015***	0.016***	0.015***	0.012***	0.014***	0.010**	0.016***
	(0.004)	(0.004)	(0.003)	(0.004)	(0.004)	(0.004)	(0.005)	(0.003)
L.GDP per capita (log)	−2.136***	−2.379***	−2.214***	−2.253***	−2.391***	−2.703***	−1.616***	−2.058***
	(0.179)	(0.177)	(0.195)	(0.188)	(0.176)	(0.197)	(0.284)	(0.226)
L.Area (log)	0.249***	0.343***	0.460***	0.335***	0.363***	0.410***	0.276***	0.362***
	(0.081)	(0.063)	(0.080)	(0.072)	(0.065)	(0.085)	(0.080)	(0.064)
Federalism			0.557***					
			(0.195)					
L.Employment				0.092				
				(1.732)				
L.Left government					−0.365***			
					(0.119)			

(*continued*)

Table 6.2 (continued)

	(1)	(2)	(3)	(4)	(5)	(6)	(7)	(8)
L.Number of government parties						0.232*** (0.073)		
L.Labor mobility							2.905 (5.831)	
L.Concentration^2								214.464** (93.778)
Constant	19.393*** (2.607)	20.646*** (2.385)	16.520*** (3.071)	19.681*** (2.908)	21.488*** (2.416)	23.302*** (2.480)	13.431*** (2.592)	17.284*** (2.736)
Observations	151	151	151	132	151	132	133	151
R-squared	0.549	0.583	0.616	0.569	0.593	0.646	0.278	0.600

From 2SLS Model. Robust standard errors appear in parentheses. All models include year fixed effects. Year coefficients are not reported due to space constraints. *** $p < 0.01$, ** $p < 0.05$, * $p < 0.1$. In the first stage model, *Open List* is predicted using historical periods, as described in the text.

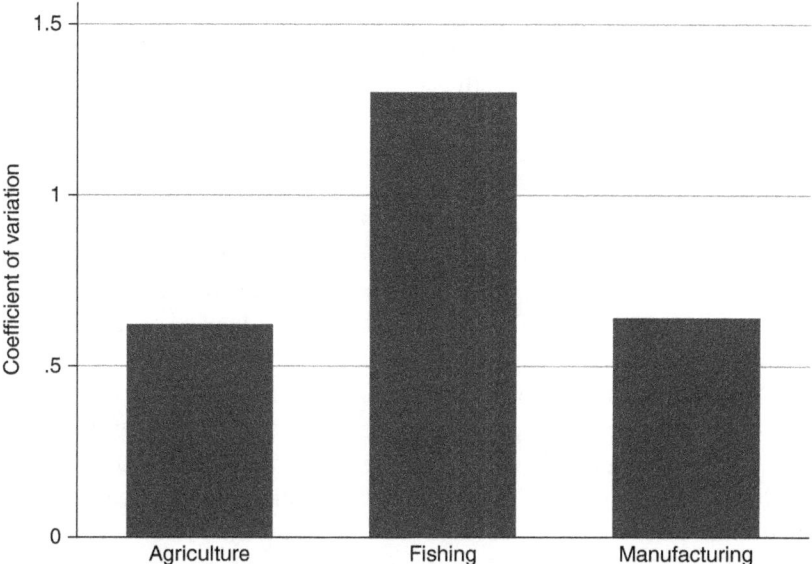

Figure 6.2 Variation in the geographic concentration of economic sector employment in Norway from 2008–2012.
Notes: Author's calculations using employment data from Statbank (www.ssb.no/en/statistikkbanken)

Arguably the instrumental variables allow us to more closely estimate the causal relationship between electoral systems and subsidy spending. Instruments may, in fact, get us as close as possible to identifying causal relationships given that we cannot randomly assign electoral rules or list systems to countries. Can we ever know if electoral rules "cause" distinct policy outcomes, given the myriad unobservable factors that drive the selection of electoral systems and policies? Ultimately, to know this we would need to answer the counterfactual question: If we picked a country at random and went back in history to change its electoral rules, how would this alter its current economic policies? (Persson and Tabellini 2003). The problem, of course, is that we cannot observe the relevant counterfactual. This daunting challenge should not lead scholars to abandon research on electoral systems and policy. Understanding how electoral institutions, a fundamental feature of democracy, influence policy outcomes is an important research agenda. And much can be learned from observational data, as demonstrated here. Although observational data do not typically identify causal effects, instrumental variables help to reveal estimates closer to causal relationships and uncover interesting and important patterns, which can feed into theory building.

A few words about the control variables are in order. Left governments spend less on subsidies than right governments, all else equal. In both Tables 6.1 and 6.2, the coefficient on *Left Government* is negatively signed and statistically significant. Among PR countries, left-leaning governments tend to allocate less of their budget to subsidies than right-leaning governments, all else equal.

Coalition governments spend more on subsidies when they included relatively more parties, all else equal. The estimated coefficient on the *Number of parties in government* is positive and statistically significant in Tables 6.1 and 6.2. The more parties involved in a government coalition, the more money allocated to subsidies, all else equal. This result confirms the idea that coalition governments typically spend more money than single-party governments because each party in government funds programs for their own supporters (Bawn and Rosenbluth 2006). Although the number of government parties influences subsidy spending, electoral systems continue to have a robust, independent effect. Even when the number of parties in government is included as a control variable, the estimated coefficient on *Open List* remains statistically significant.

Within-Country Test

Subsidies vary within – not just between – countries. Even in countries with generous subsidies, not all sectors benefit equally from government support. Some sectors receive lavish amounts of state aid while others receive relatively little financial assistance from the government. How can the variation in subsidies *within* a country be explained?

National electoral institutions are constant within a country – they typically do not vary over the short to medium term. On their own then electoral institutions cannot explain the varied generosity of subsidies between different economic sectors within a given country. Instead, the within country variation in subsidies is explained, in part, by economic geography. In any given country, employment in some sectors will be more geographically concentrated than in other sectors. I exploit the cross-sectoral variation in employment patterns in a single country to investigate the policy effects of economic geography. A single-country study holds constant electoral institutions and other time-invariant, country-specific factors, such as culture. Holding institutions constant isolates the effects of economic geography on policy outcomes. However, electoral institutions interact with economic geography to shape economic policy. So even when electoral institutions are held constant, as in a single-country case study, it is important to understand the country's electoral system to understand how economic geography will shape economic policy.

The country under investigation here is Norway. Norway has a proportional representation system with de facto closed party lists. Voters cast a ballot for a party list and the names on a party's list correspond with the candidates representing that party. The candidates are chosen by the nomination conventions of each party (Sørensen 2003). In theory, voters may modify the order of candidates on the list. Voters are allowed to change the rank order of the candidates on the party list as well as cross out candidates (Aardal, 2011: 8). However, the levels of coordination required to overturn the parties' rankings are so extreme that they effectively deter attempts to do so. At least half the voters have to make exactly the same alterations of the list for it to have any effect (Aardal, 2011: 8). For all practical purposes, Norway's system is effectively a closed-list system (Aardal, 2011: 8).

How does economic geography shape policy in a closed-list PR system like Norway? Geographically diffuse groups are politically advantaged in closed-list PR systems. As a result, diffuse groups win greater subsidies in closed-list systems as compared to open-list systems, as illustrated by the cross-national results reported in Tables 6.1 and 6.2. Diffuse groups win more subsidies in closed-list systems because closed lists engender party-centered electoral competition, generate high levels of party discipline, and align the incentives of individual legislators and party leaders. In closed-list systems, individual legislators have few incentives or opportunities to champion the interests of geographically concentrated groups. As a result, concentrated groups often find themselves without a champion at the national level in closed-list systems. Parties and their legislators instead work to assist groups spread across the country to maximize the party's legislative seats. Sectors with geographically diffuse employment will consequently win more generous subsidies than concentrated sectors in closed-list PR systems.

I test this hypothesis using novel data on government-funded subsidies to various sectors in Norway. The Ministry of Finance provided the subsidy data to me upon request. The data indicate the net subsidy costs in millions of Norwegian krone (NOK) at 2013 prices deflated by annual inflation for the gross domestic product of mainland Norway.[14] Using these data, I calculate the amount spent on subsidies for three economic sectors that exhibit varying degrees of geographic concentration: agriculture, fishing, and manufacturing. My first dependent variable

---

[14] Lone Semmingsen, Deputy Director General, Ministry of Finance, Norway, Written communication, January 20, 2015.

equals the amount spent on subsidies to each of these sectors. My second dependent variable equals the amount spent on subsidies to a given sector as a percentage of total subsidy spending. The second measure reports how much of the government's total subsidy budget went to a particular sector of the economy. I hypothesize that more diffuse sectors will win more of the government's total subsidy budget in this de facto closed-list PR system, controlling for the size of the sector.

Measuring Geography Concentration

To measure the geography of sectors' employment, I use highly disaggregated employment data from Statistics Norway – the national statistical institute of Norway and the main producer of official economic statistics.[15] These data report the number of employed persons by sector, electoral district, and year. I calculate the standard deviation of each sector's employment for each of Norway's nineteen electoral districts for every each year during the period from 2008 to 2012. The standard deviation of a sector's employment across electoral districts provides politically relevant information about the geographic distribution of employees. If all of a sector's employees were located in a single electoral district, the standard deviation would equal one – its maximum possible value. The standard deviation of sector employment across electoral districts provides a measure of geographical concentration that corresponds with the theoretical concept of interest and varies within a single country.

As an alternative measure of geographic concentration, I calculate the coefficient of variation (COV), which equals the standard deviation of sector employment divided by the mean. The COV shows the extent of variability in relation to the mean of the population. In other words, the coefficient of variation normalizes the standard deviation with respect to the mean, which is useful when the means vary. Figure 6.2 reports the COV for three important sectors of the Norwegian economy averaged over the period from 2008 to 2012.[16]

The geography of employment varies in between sectors, as illustrated in Figure 6.2. Of the three sectors investigated here, agriculture employment is the most geographically diffuse. Agriculture employs a relatively similar number of people in all nineteen of Norway's electoral districts. Given this, agriculture is characterized as being a geographically diffuse sector in

---

[15] www.ssb.no/en/statistikkbanken.
[16] This measure of geographic concentration differs from the measure used in Chapter 4 in order to maximize the number of sectors included in the single-country sample.

Norway. In fact, Norway's agriculture sector is more geographically diffuse than the OECD average (OECD 2008).

In contrast, employment in the Norwegian fishing sector was geographically concentrated during the period from 2008 to 2012.[17] Today, more than half of the fishing sector's labor force lives in just three electoral districts: Finnmark, Nordland, and Møre og Romsdal (Fløysand and Jakobsen 1999). Traditionally, employment in the fishing industry was more evenly spread across the country. However, in recent years the industry has become more and more concentrated (Fløysand and Jakobsen 1999). The industry's increased geographic concentration may explain why political support for fishing subsidies has fallen. Over the period from 1980 to 1990, spending on fishing subsidies fell by around 90 percent (Ásgeirsdóttir 2008). In 1981, the government spent 133 million krone on support for the fishing industry (Isaksen 2000). These support measures included minimum income guarantees, bait subsidies, insurance subsidies, and price support programs (Isaksen 2000, Schrank 2003). By 1999, spending on fishing-industry subsidies fell to just 86 million krone (Isaksen 2000).

The Norwegian government was able to withdraw financial support from the fishing industry without incurring major political costs (Ásgeirsdóttir 2008). The political risks from cutting fishing subsidies fell as the industry's employees became more and more geographically concentrated. The industry's remaining employees are almost entirely concentrated along the coast in just three electoral districts (Fløysand and Jakobsen 1999). The industry's increased geographical concentration translated into less political clout in Norway's closed-list PR system, as made clear in interviews with members of the union and employer association for Norwegian fisherman (*Norges Fiskarlag*). One interviewee said, "We (still) have influence, but we feel we have been sidelined" (Ásgeirsdóttir, 2008: 70). Changes in the industry's employment patterns left, "fewer voters for the fisheries to mobilize" (Ásgeirsdóttir, 2008: 70) and as a result, the industry enjoys less political influence that it did when its employees were geographically diffuse.

The fishing industry's experience provides anecdotal evidence of the political importance of economic geography in closed-list PR systems. I aim to provide systematic evidence of this pattern using an econometric model of government spending on sector-specific subsidies. To this end, I regress sector-specific subsidies on sector employment patterns for three sectors of the Norwegian economy (agriculture,

---

[17] The OECD report combines agriculture and fishing and as a result it obscures the variation in concentration in these two sectors.

Table 6.3 *Effect of geographic concentration on sector-specific subsidies in a closed-list PR system*

	(1) Subsidies (NOK mil)	(2) Percent of total subsidies	(3) Subsidies (NOK mil)	(4) Percent of total subsidies
Coefficient of variation	−15.137*** (2.494)	−0.666*** (0.110)		
Standard deviation of sector employment			−25.265*** (2.140)	−1.111*** (0.094)
Total sector employment			0.772*** (0.065)	0.034*** (0.003)
Constant	20.396*** (2.310)	0.897*** (0.102)	14.673*** (0.969)	0.645*** (0.043)
Observations	15	15	15	15
R-squared	0.739	0.739	0.922	0.921

Notes: The unit of analysis is sector-year. Standard errors appear in parentheses. *** p<0.01, ** p<0.05, * p<0.1.

fishing, and manufacturing) over the course of five years (2008–2012). The sample size is limited by data availability. Given the small sample, I estimate a parsimonious model with a few key control variables. The results, reported in Table 6.3, must consequently be treated with caution.

Within Country Results

Governments spend more money on subsidies for geographically diffuse sectors than on geographically concentrated sectors in Norway, a de facto closed-list PR country. Sectors that employ a comparatively uniform number of people across the country's nineteen electoral districts win relatively more generous subsidies. The greater the variance in a sector's employment between electoral districts, the fewer krone the sector receives in government-funded subsidies.

The estimated coefficients on both measures of geographic concentration are negatively signed and statistically significant in all models. The coefficient of variation (COV) is negatively correlated with subsidies, as reported in columns 1 and 2 of Table 6.3. An increase in the COV by one standard deviation over its mean value results in a 53.6 percent decrease in subsidies. In other words, sector subsidies fall from 7.44 million NOK to 3.45 million NOK when the coefficient of variation increases by one standard deviation over its mean value. An increase in the standard deviation of sector employment also correlates negatively with subsidy spending. In sum, sectors with more geographically concentrated employment win relatively fewer subsidies in this closed-list PR country.

Importantly, the negative correlation between sector concentration and sector subsidies remains statistically significant when controlling for a sector's total employment. Larger sectors in terms of employees receive more subsidies than smaller sectors, as illustrated in columns 3 and 4 of Table 6.3. However, the geographic diffusion of employees remains a robust predictor of subsidies controlling for total employment. This is an important finding. It is often difficult to isolate the impact of total employment from the geographic concentration of employment, as the fishing industry example suggests. Employment in Norway's fishing industry declined and became increasingly concentrated at the same time, making it difficult to isolate the policy effects of these two simultaneous changes. However, in the econometric model, it is possible to control for total employment and consequently isolate the effects of geographic concentration on subsidy spending. Holding total employment constant, increased geographic concentration reduces government spending on subsidies in closed-list PR systems.

DISTRICT MAGNITUDE AND LIST TYPE

The type of list system used in countries with proportional electoral systems influences the generosity of government subsidies. The effect of a list system on governments' subsidy spending depends on the geographic distribution of beneficiaries. In this way, list systems interact with economic geography to influence subsidy spending. List systems may also interact with district magnitude to shape politicians' incentives to supply certain economic policies, such as subsidies. District magnitude refers to the number of candidates elected in a district.[18] District magnitude varies among PR countries.[19] In some countries, as few as two legislators are elected from each district, as in Chile. In others, as many as 150 legislators are elected from a single district, as in Slovakia where the entire country makes up one electoral district.

---

[18] District magnitude is distinct from the geographical size of a district and the number of voters in it.
[19] While district magnitude typically equals one in plurality systems, it varies when comparing proportional systems.

District magnitude may interact with a country's list type to shape politicians' incentives to cultivate a personal vote (Carey and Shugart 1995, Shugart, Valdini, and Suominen 2005, Chang and Golden 2007, Carey and Hix 2011). Increases in district magnitude intensify personal vote incentives in systems where institutions, such as open lists, already encourage personal vote-seeking (Carey and Shugart 1995). As district magnitude increases, the number of co-partisans from which a given candidate must distinguish herself grows. As a result, the importance of establishing a unique personal reputation, distinct from that of the party, increases with district magnitude in open-list PR systems (Carey and Shugart 1995).

Politicians can build their personal support bases using subsidies. However, the usefulness of subsidies to develop a personal vote depends on the geographic concentration of the beneficiaries. When beneficiaries are geographically concentrated in a legislator's district or bailiwick, subsidies can be an effective means by which to cultivate a personal vote. In contrast, when beneficiaries are geographically diffuse across the country, promising subsidies to that sector will be an inefficient way to cultivate a personal vote. Given the powerful incentives that exist to cultivate a personal vote in open-list PR systems when district magnitude increases, I hypothesize that subsidies will be relatively more generous when beneficiaries are geographically concentrated and district magnitude is relatively high.

I test this hypothesis by estimating the effects of district magnitude and economic geography on subsidies in a sample of open-list PR countries. To measure district magnitude, I use the average number of representatives elected to the lower (or only) legislative chamber by each district. I interact this measure of mean district magnitude with geographic concentration. The results are reported in Table 6.4 and displayed graphically in Figure 6.3.

In open-list PR systems, the marginal effect of mean district magnitude on subsidy budget shares is positive and statistically significant when beneficiaries are geographically concentrated. Whenever geographic concentration is greater than 0.043, mean district magnitude is positively correlated with subsidy spending. Ten percent of the sample observations fall in this range. This result is consistent with the argument that higher district magnitude reinforces politicians' incentive to cultivate a personal vote in systems that already encourage personal vote-seeking, such as open-list PR.

Politicians competing in open-list PR systems fund subsidies when the beneficiaries are geographically concentrated in order to cultivate a personal vote. When mean district magnitude is at its highest observed

Table 6.4 *Effect of mean district magnitude on subsidies in open-list PR*

	(1)	(2)	(3)	(4)	(5)	(6)	(7)	(8)
L.Mean District Magnitude (log)	-0.170***	-1.029***	-0.754**	-0.616*	-1.008***	-1.069***	-0.562	-0.988***
	(0.064)	(0.300)	(0.291)	(0.319)	(0.296)	(0.288)	(0.424)	(0.324)
L.Concentration	-29.463***	-84.793***	-67.474***	-58.196***	-84.571***	-88.887***	-58.384**	-91.390***
	(3.190)	(16.820)	(16.006)	(18.298)	(16.438)	(15.523)	(23.807)	(17.713)
L.MDM*L. Concentration		30.014***	20.689**	18.417*	29.503***	31.507***	15.745	29.876***
		(9.570)	(9.150)	(9.561)	(9.386)	(8.933)	(12.525)	(9.634)
L.Trade	-0.000	-0.008*	-0.005	-0.010**	-0.009**	-0.009**	-0.006	-0.008**
	(0.004)	(0.004)	(0.004)	(0.004)	(0.004)	(0.004)	(0.005)	(0.004)
L.GDP per capita (log)	-2.857***	-2.670***	-2.966***	-3.275***	-2.662***	-2.750***	-2.946***	-2.625***
	(0.306)	(0.315)	(0.301)	(0.397)	(0.313)	(0.310)	(0.630)	(0.342)
L.Area (log)	0.270***	0.178*	0.259***	0.067	0.195*	0.184*	0.241**	0.192*
	(0.090)	(0.094)	(0.087)	(0.143)	(0.106)	(0.109)	(0.096)	(0.106)
Federalism			0.692***					
			(0.144)					
L.Employment				7.591*				
				(4.500)				

(*continued*)

Table 6.4 (*continued*)

	(1)	(2)	(3)	(4)	(5)	(6)	(7)	(8)
L.Left government					-0.149			
					(0.187)			
L.Number of government parties						0.107		
						(0.084)		
L.Labor Mobility							-2.325	
							(7.872)	
L.Concentration^2								76.568
								(127.828)
Constant	28.078***	29.671***	29.714***	33.928***	29.130***	29.987***	29.550***	29.055***
	(3.763)	(3.577)	(3.544)	(4.683)	(3.573)	(3.381)	(6.726)	(4.111)
Observations	95	95	95	82	95	82	77	95
R-squared	0.817	0.831	0.848	0.815	0.834	0.850	0.601	0.832

Notes: Robust standard errors appear in parentheses. All models include year fixed effects. Year coefficients are not reported due to space constraints. *** $p<0.01$, ** $p<0.05$, * $p<0.1$.

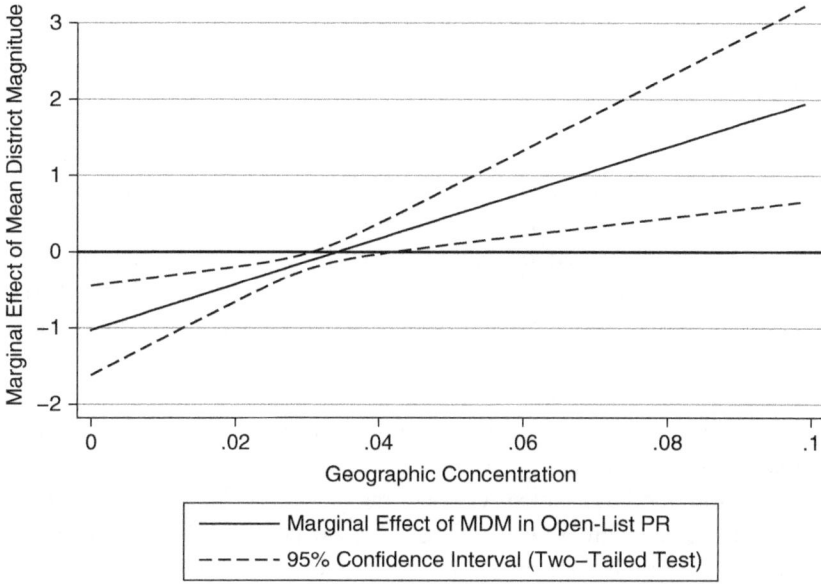

Figure 6.3 Marginal effect of mean district magnitude on subsidy budget shares in open-list PR

value in the sample of open-list PR countries, increasing *Concentration* from its lowest observed value to its highest value increases subsidy spending by nearly seven percentage points.[20] This result suggests that the usefulness of subsidies for developing a personal vote depends on the geographic concentration of subsidy recipients. Only when beneficiaries are sufficiently geographically concentrated are subsidies an effective means by which to cultivate a personal vote.

When beneficiaries are geographically diffuse, increases in mean district magnitude do not correlate with greater subsidy spending. In fact, mean district magnitude is negatively correlated with subsidies in open-list PR countries when the beneficiaries are geographically diffuse. When *Concentration* equals its' lowest possible value (i.e. zero), a one-unit increase in the natural log of mean district magnitude reduces subsidy spending by one percentage point, all else equal. The negative marginal effect of mean district magnitude on subsidies is statistically significant at the 95 percent level whenever *Concentration* is less than 0.03, which holds for 73 percent of the sample. The negative effect of mean district magnitude on subsidies at low levels of geographic concentration suggests that subsidies are an ineffective tool with which to cultivate

[20] Calculated using column 2 from Table 6.3.

a personal vote when the beneficiaries are geographically diffuse. Legislators consequently prioritize other spending areas.

In Chapter 3, I identified personal vote seeking as a potential mechanism linking electoral institutions and policy outcomes. The results reported here provide empirical support for this mechanism. A positive correlation exists between district magnitude and subsidy spending when beneficiaries are geographically concentrated in open-list PR systems and district magnitude is high. This correlation arguably exists because politicians competing in open-list systems with high district magnitude work harder to cultivate a personal vote and do so using subsidies when the beneficiaries are concentrated geographically.

If personal vote seeking is indeed the mechanism at work here, I would expect to find no correlation between district magnitude and subsidies in closed-list PR systems. In closed-list systems, increased district magnitude should not increase subsidy spending because legislators have no incentives to cultivate a personal vote – regardless of the number of seats up for grabs in their district. Therefore, district magnitude should have no effect on subsidy spending in closed-list systems. To test this possibility, I reestimate the models using only closed-list countries. The results for closed-list countries, reported in Table 6.5, show no robust correlation between district magnitude and subsidies in PR systems with closed-party lists (or the interaction of district magnitude and geography). In closed-list systems, increases in district magnitude do not correlate with greater government spending on subsidies, all else equal.[21]

Similarly, in a PR country with de facto closed party lists (Norway), I find no robust correlation between district magnitude and subsidy spending (see Chapter 7). In Norway, district magnitude ranges from 4 to 19. The variation in district magnitude within Norway does not influence the geographic distribution of subsidies across the country. Norwegian electoral districts with more representatives receive no more generous subsidies than districts with fewer representatives. The fact that district magnitude does not correlate with subsidies in closed-list PR systems, at any level of geographic concentration, suggests that personal vote seeking does indeed influence government spending on industrial subsidies.

---

[21] In Table 6.5, two of the coefficients on *Mean District Magnitude* are statistically significant at conventions levels. However, the first is in the model that excludes the key interaction term (Concentration*MDM) and is negatively signed. The second is in the model that includes the squared *Concentration* term for which the estimated effects are unreasonably large, presumably due to the relatively small sample size (n = 54).

Table 6.5 Effect of mean district magnitude on subsidies in closed-list PR

	(1)	(2)	(3)	(4)	(5)	(6)	(7)	(8)
L.Mean District Magnitude (log)	-1.134*	3.566	-6.172	3.781	4.482	4.169	2.852	10.061***
	(0.644)	(3.476)	(4.875)	(3.767)	(2.910)	(3.582)	(3.458)	(3.372)
L.Concentration	-47.564	214.9	-102.4	201.9	243.3	213.8	185.8	771.1***
	(28.2)	(190.2)	(211.6)	(199.5)	(158.3)	(197.3)	(189.2)	(243.8)
L.MDM*L. Concentration		-99.21	57.86	-100.1	-116.8*	-107.4	-85.7	-206.0***
		(71.48)	(87.65)	(75.43)	(60.87)	(72.57)	(71.15)	(66.63)
L.Trade	0.014	0.028	0.047*	0.056**	0.037	0.053**	0.022	0.038
	(0.022)	(0.024)	(0.024)	(0.026)	(0.024)	(0.024)	(0.023)	(0.022)
L.GDP per capita (log)	-1.770	-0.035	3.848*	-0.388	-0.614	-1.282	0.037	0.880
	(1.456)	(1.610)	(1.926)	(2.112)	(1.481)	(1.894)	(1.587)	(1.133)
L.Area (log)	-0.056	0.269	7.170**	1.103	0.574	0.876	0.078	0.805
	(0.485)	(0.515)	(2.683)	(0.688)	(0.543)	(0.532)	(0.503)	(0.516)
Federalism			15.138**					
			(5.698)					
L.Employment				9.025				
				(16.183)				
L.Left government					-0.554*			
					(0.309)			

(continued)

Table 6.5 (*continued*)

	(1)	(2)	(3)	(4)	(5)	(6)	(7)	(8)
L.Number of government parties						-0.043		
						(0.174)		
L.Mobility							15.695*	
							(8.340)	
L.Concentration^2								-2724.3**
								(1099.5)
Constant	22.498**	-12.024	-117.8**	-22.129	-12.122	-9.304	-8.610	-50.801**
	(10.540)	(25.771)	(45.012)	(30.373)	(21.293)	(28.427)	(25.534)	(20.900)
Observations	54	54	54	48	54	48	54	54
R-squared	0.594	0.636	0.684	0.704	0.676	0.702	0.660	0.701

Notes: Robust standard errors appear in parentheses. All models include year fixed effects. Year coefficients are not reported due to space constraints.
*** p<0.01, ** p<0.05, * p<0.1.

CONCLUSION

The simple distinction between plurality and proportional electoral systems together with economic geography helps to explain the cross-national variation in governments' particularistic economic policies. However, it fails to explicate the variation in subsidy spending among countries with proportional systems. Governments elected via proportional rules in different countries spend varied amounts on subsidies. Why? I offer two concordant explanations. First, the policy effects of electoral institutions depend on economic geography. Because economic geography varies between countries, similar institutions can have different policy effects. Second, key institutions vary among countries with proportional systems, including list type and district magnitude. Holding economic geography constant, list type and district magnitude influence government spending on subsidies in PR countries, as I demonstrate in this chapter.

Although many aspects of countries' electoral systems vary, this variation is masked by the blunt distinction between plurality and proportional systems. Some PR systems have characteristics more similar to plurality systems than to other PR systems. Elections in open-list PR systems, for example, are candidate-centered. Candidates on the same-party list compete against one another for "preference votes" in multimember districts. Politicians work to cultivate a personal vote to maximize their chances of winning office. In this way, electoral competition in open-list PR systems looks similar to electoral competition in plurality systems (although in plurality systems co-partisans typically compete against one another in primaries). This observation raises doubts about the usefulness of the plurality/PR distinction. The distinction overlooks a key characteristic shared by both plurality systems and open-list PR systems: candidate-centered electoral competition. More generally, the blunt dichotomy between plurality and proportional obscures important institutional variation within each system. Moving beyond this blunt distinction can help to improve our understanding of the effects of electoral institutions on policy outcomes, as illustrated in this chapter.

7

The Policy Effects of Electoral Competitiveness in Closed-List PR

Government-funded subsidies vary between countries. However, they also vary within countries. Governments in a given country often spend more on subsidies for some economic sectors than others. Within a country, the variation in subsidies between sectors can be partly explained by economic geography. Employees in some sectors are more geographically concentrated than others and the geographic distribution of a sector's employees, together with a country's electoral system, influences government spending on subsidies for the sector. Governments in closed-list PR systems spend more on geographically diffuse sectors than on concentrated sectors, as illustrated in Chapter 6. In this way, economic geography helps to explain the variation in government-funded subsidies between sectors within countries.

While the generosity of government subsidies varies between sectors within countries, it also vary between regions. Governments frequently spend more on subsidies to some regions than others. In Norway, for example, the government spent fifteen times more money on subsidies to producers in the northern region of Troms than the western region of Rogaland. Similarly, the government of Belgium spent €2,600 per person on subsidies to one canton but spent absolutely nothing on subsidies for another canton.[1] Also, in France, the government funded subsidies for wine makers in the Cognac region worth €1,524 per hector but declined to make these subsidies available to producers in other regions (see Chapter 5).

Subsidies vary between regions within countries even controlling for the geographic distribution of economic activity, as demonstrated by the amount spent on subsidies per employee. In 2012, for example, the Norwegian government spent 309 krone per manufacturing sector

---

[1] In the year 2008.

employee in the southern region of Vestfold. The government spent eighteen times more per manufacturing sector employee that same year in the central region of Oppland where manufacturing subsidies equaled 5,523 krone per employee. As this example illustrates, subsidy spending can vary radically between different regions in a country, even after controlling for the geographic distribution of employees. This observation raises an important question: Why do governments spend more money on subsidies for some regions than others controlling for employment patterns?

I argue that some regions receive more generous subsides than others because of governments' reelection incentives. Governments spend more money on subsidies for certain areas to maximize their reelection prospects. In order to consolidate the electoral advantage that helped them win office in the first place, parties in government provide more generous subsidies to some electoral districts, depending on the competitiveness of elections in those districts.[2]

Although the effects of district-level electoral competitiveness have been studied extensively in plurality countries, scant attention has been paid to the possibility that parties respond to variations in district-level electoral competitiveness in proportional systems.[3] Conventional wisdom suggests that parties in PR systems have few incentives to target benefits to select districts because all votes are equally valuable.[4] Every vote does, in fact, contribute to a party's electoral success in PR systems. And when a single national district is used to elect a country's legislators, all votes are equally valuable – regardless of their geographic location. But most PR systems have more than one electoral district and when parties compete in multiple districts, some votes will be more valuable than others. This fact

---

[2] Although electoral competitiveness influences the distribution of subsidies within a country, it is less useful for explaining the cross-national variation in subsidies. Competitiveness varies between districts within countries. Elections in some districts are more competitive than others, and as a result it is difficult to construct a theoretically relevant country-level measure of electoral competitiveness. Additionally, measures of competitiveness must capture how the concept varies across different electoral systems. For these reasons, district-level electoral competitiveness, while useful for explaining the variation in subsidies between districts within a country, is less useful for explaining the variation in subsidies between countries.

[3] For example, McGillivray's empirical tests include only plurality countries, notably the US and Canada. Her tests do not extend to PR countries (McGillivray, 1997: 271, McGillivray, 2004: 81). McGillivray herself writes, "The hypotheses for proportional representation systems are not examined" (2004: 87).

[4] However, individual legislators work to divert money to groups concentrated in their own districts in open-list PR systems in order to cultivate a personal vote (see Chapter 6 and Golden and Picci 2008).

raises the possibility that parties may target benefits to select districts for electoral gain – even in PR systems.

In this chapter, I investigate the distribution of subsidies between electoral districts in an archetypal PR country: Norway. Like many PR countries, Norway lacks the institutional attributes usually associated with policy targeting or pork-barrel politics. Norway has a political system believed to be highly resistant to particularistic policies (Tavits 2009) – namely a parliamentary system, with strong parties and party-centered elections. Given this, Norway presents a "least likely" case for geographic targeting. In fact, few scholars expect to see policy targeting in a political system like Norway's (Shugart 1999, Denemark 2000, Crisp et al. 2004, Morgenstern and Swindle 2005). Most scholars focus instead on national level policies when studying the electoral strategies of incumbent governments in PR systems (Alesina, Roubini, and Cohen, 1997).

I examine the possibility that political parties in proportional systems use local goods, such as targeted subsidies, to win reelection. To do so, I use novel subsidy data to calculate government spending on manufacturing subsidies per manufacturing-sector employee in each of Norway's nineteen electoral districts. I find that subsidy spending per employee is higher in districts where the largest government party won a greater share of the votes in the previous election, all else equal. This result suggests that parties competing in closed-list PR systems, like Norway, target economic benefits to "safe" districts.

Recall that in Chapter 6, I reported evidence that governments in Norway spend more money on geographically diffuse sectors than on concentrated sectors. How can governments spend more money on diffuse sectors and target subsidies selectively at the same time? The two strategies are not mutually exclusive. In closed-list PR systems, parties' first-best strategy is to fund subsidies for sectors whose distribution of employees closely matches the geographic distribution of the party's supporters. Parties that manage to become the largest party in government in PR systems will tend to have geographically diffuse support. Parties with supporters in only a few geographically concentrated locations are unlikely to become the largest government party in a multiparty proportional system. As a result, the distribution of employees in diffuse sectors is more likely to match the geographic distribution of the party's supporters than concentrated sectors. The party's first best strategy is to target sectors with geographically diffuse employment. If, however, a sector's employees are imperfectly distributed relative to a party's supporters, the party's second best option is to target subsidies to safe districts. Diffuse sectors provide parties with the widest range of possible options for geographic-targeting. A sector that employs people across the entire country allows parties to selectively target benefits to any district via more (or less) generous subsidies. Government parties in closed-list PR systems will therefore spend relatively more money on diffuse sectors, as compared to concentrated sectors, and subsidy spending per person will tend to be higher in "safe" districts, as I demonstrate using novel quantitative data on government-funded subsidies.

I supplement the quantitative results with qualitative evidence obtained from interviews with government ministers and bureaucrats responsible for subsidy programs in Norway. These interviews confirm the importance of electoral politics and economic geography for governments' spending decisions. The interviews also illustrate the mechanisms that government parties use to target subsidies to politically important areas. Both the quantitative and qualitative evidence from this single-country case study confirm the importance of electoral incentives for economic policy making in democratic countries.

EXPLAINING WITHIN-COUNTRY VARIATION

Within a given country, producers in some electoral districts receive more generous economic support from the government than others. In Norway, for example, government spending on subsidies is greater in some electoral districts than others. The variation in subsidies between Norway's electoral districts is illustrated in Figure 7.1, which reports average government spending on manufacturing subsidies per manufacturing sector employee from 2005 to 2012 for Norway's 19 electoral districts. Figure 7.1 illustrates the key question motivating this chapter: why do subsidy amounts vary per person between electoral districts within a given country?

I hypothesize that the competitiveness of elections influences the generosity of government subsidies. The competitiveness of elections often varies between districts in democratic countries with multiple electoral districts.[5] Some districts may be relatively "safe" for a given

---

[5] Competitiveness is often defined by a legislator's margin of victory (Fiorina 1973). In a single-member district, the margin of victory is easy to calculate; it simply equals the number of votes between the first and second place finishers. In multimember districts, calculating any individual legislator's "margin of victory" is far more difficult. Other scholars define competitiveness for different units of analysis. For example, Kayser and Lindstädt (2015) define competitiveness as the expected probability that the plurality party in parliament loses its seat plurality in the next election (p. 243). See Strøm (1992) for a theoretical definition of competitiveness.

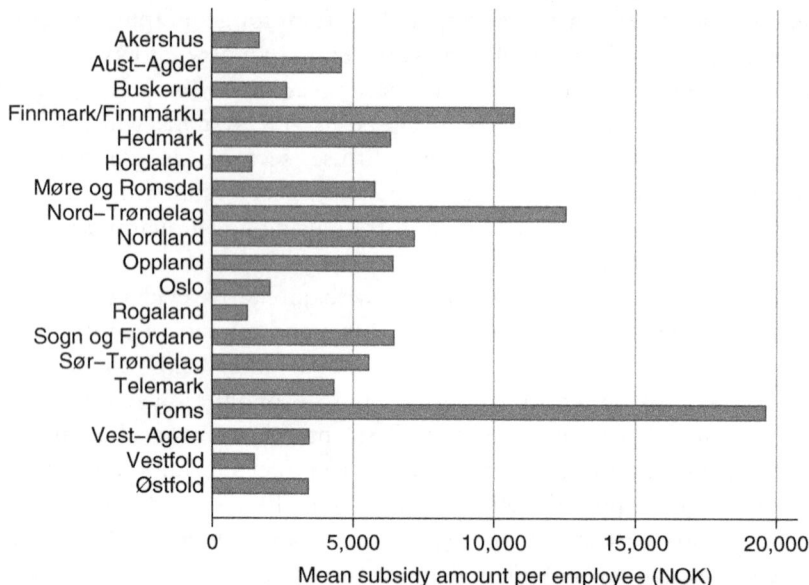

Figure 7.1 Average subsidy amount per manufacturing employee, 2005–2012
Source: Author's calculations from data provided by Innovation Norway. All amounts reported in Norwegian krone (NOK).

party – that is, a particular party may tend to win a large share of the district's votes. In contrast, other districts may be more competitive for a given party – that is, a party may run neck-and-neck with another party in a given district. Parties generally know how competitive they are in a given district. In PR systems, parties know this by simply observing the number of legislative seats they win in each district because the allocation of legislative seats is proportional to parties' district vote share. In some proportional systems, parties also have access to the protocols for the last election that describe the precise calculations for how votes are translated into seats and report parties' vote shares by district. This is the case in Norway, for example.[6]

Variation in district-level electoral competitiveness may influence the geographic distribution of subsidies within a country.[7] A large body of scholarship argues that district-level competitiveness shapes the geographic distribution of economic rents. Such arguments have been developed almost exclusively in the context of plurality electoral systems

---

[6] See www.stortinget.no/globalassets/pdf/innstillinger/stortinget/2013–2014/inns-2013 14-001.pdf.
[7] However, it is not clear what, if any, implication this observation may have for the cross-national variation in subsidy spending.

with single-member districts. A debate exists over precisely how district-level electoral competitiveness matters for policy targeting in plurality systems. One side of this debate posits that rents targeted to competitive districts bring about greater electoral rewards than rents targeted to "safe" districts.[8] In competitive districts, politicians work to earn every available vote because each further vote is electorally valuable in a tight race. To this end, incumbents will seek to influence the geographic allocation of government assistance in competitive districts. Directing benefits to their own districts can increase their chances of winning office by securing additional votes. In contrast, politicians who command a large margin in "safe" districts feel less need to chase after each additional vote because additional votes do not increase their chances of winning office. Instead, they simply add to an already large margin. As a result, incumbents in safe districts have fewer incentives to work to secure economic benefits for their constituents. In France, for example, legislators who won by larger margins were less likely to lobby for subsidies for their wine-making constituents (see Chapter 5). Given this pattern, subsidies and other economic incentives may go disproportionality to competitive districts in plurality countries. But what role, if any, does competitiveness play in countries with proportional electoral rules?

Most countries around the world today use some form of proportional system, yet the effects of electoral competitiveness in PR systems remain largely unknown. Two factors account for the lack of attention to electoral competitiveness in PR systems. First, the difficulty of identifying "competitive" districts in multimember, multiparty PR systems makes empirical research on the topic challenging. Second, theoretical models typically assume – either explicitly or implicitly – that only one nationwide electoral district exists in PR systems (e.g. Persson and Tabellini 2003, McGillivray 2004). Grossman and Helpman (2005), for example, model a proportional system with a single nationwide legislative constituency. If there is only one, nationwide electoral district, then by definition no within country variation exists in electoral competitiveness. Yet, few real world PR systems have just one nationwide district. Most PR countries have multiple, geographically defined electoral districts that encompass subsegments of the country. In PR countries with multiple districts, most legislative seats are awarded to parties based on their share of a district's votes – rather than their share of the national vote. The district-level allocation of seats raises the possibility that some districts are relatively more competitive for certain parties than others. District-level electoral

---

[8] In contrast, Cox and McCubbins (1986) argue that parties reward loyal voters.

competitiveness may be an important, yet previously overlooked, feature of PR systems with multiple electoral districts. I examine this possibility by investigating the impact of district-level electoral competitiveness on particularistic economic policies in an archetypal PR country: Norway.

WHY NORWAY?

Norway provides a valuable case study for several reasons.[9] First, Norway uses proportional electoral rules to elect members of parliament. To date, nearly all research on electoral competitiveness has been conducted in plurality countries, most notably the United States. As a result, little is known about how electoral competitiveness shapes politics or policy in PR countries. Norway provides a useful case with which to explore the effects of competitiveness in PR systems.

Second, Norway is a "least likely" case for particularistic economic policies because it lacks the institutional attributes usually associated with pork-barrel politics. Norway has a parliamentary system, with strong parties and party-centered elections (Tavits 2009). Few scholars expect policy targeting in this context (Shugart 1999, Denmark 2000, Crisp et al. 2004, Morgenstern and Swindle 2005).

Third, Norway has a long history with subsidies. State subsidies date back to at least the mid-19th century, when *Kongeriget Norges Hypotekbank* was established in 1852 as a mortgage bank to provide assistance to industry (Innovation Norway 2014b). The bank granted businesses cheap loans in exchange for mortgages on property. The objective was to modernize agriculture and develop new industries. State support for business continues today. In recent years, Norway spent more on subsidies as a percentage of GDP than any other country in the European Free Trade Association (EFTA).[10]

---

[9] Alt et al. (1999) examine lobbying by firms for subsidies in Norway. However, they do not examine the distribution of subsidies between electoral districts. In fact, they focus exclusively on the demand side of the story. They measure, for example, the number of times a business organization meets with members of parliament. They point out that, "a complete model ... should also include the(se) incentives of the government-Parliament to respond [to firm lobbying] (p. 115). They concede that they "have not completely modelled the institutional supply side of policy" (p. 115). They state explicitly that they do "not know whether it helps a firm to be from a [particular] district" (p. 115). My research fills in the supply side of the story and investigates whether firms from more electorally competitive districts fare better or worse than firms from less competitive districts, all else equal.

[10] Farmers and fishers have long been assisted by the Norwegian government. This tradition of state support is increasingly being extended to technology and environmental sectors (Innovation Norway 2014b).

Finally, Norway is one of the few countries where data on government subsidies are available at the level of disaggregation needed to assess the geographic distribution of subsidies between electoral districts. Most governments are unwilling to provide detailed information on the amount of subsidies they award and to whom (Buts et al. 2012). In contrast, the Norwegian government generously provided me access to detailed subsidy data, which include the subsidy amount as well as the sector and geographic location of the recipients. These data allow for a novel investigation of the geographic distribution of state subsidies within a given country.

ELECTIONS IN NORWAY

Although national electoral institutions are constant within a country at any given point of time, they set the stage for the dynamics that play out during election campaigns and shape both the electoral incentives of political parties and their optimal (re)election strategy. For these reasons, I briefly describe Norway's electoral institutions before exploring parties' incentives to target economic benefits.

Norway has multiple electoral districts. Norway's nineteen districts correspond with the administrative provinces (*fylker*) and include the municipal authority of Oslo, which is a *fylker* in its own right. District magnitude ranges from four seats in *Aust-Agder* and *Sogn og Fjordane* to nineteen seats in Oslo. The number of seats in each district is a function of the number of citizens in a district and its geographical size (Aardal 2011).

The Norwegian Parliament, known as the *Storting*, contains 169 members that are directly elected by universal adult suffrage for a fixed term of four years. Legislators are elected via a two-tier system. One hundred fifty seats are distributed at the provincial (i.e. district) level. In other words, most legislative seats are awarded to parties by district in proportion to their share of district votes. The remaining nineteen seats are distributed as "compensatory seats" based on parties' share of the national vote. There is no formal threshold in each district, but in order to be eligible for a compensatory seat a party needs to win at least four percent of the nationwide vote.

Compensatory seats are intended to achieve a greater degree of proportionality in the overall distribution of legislative seats (Sørensen 2003). If a political party fared worse in the provincial distribution of seats than it would if the entire country had been organized as one electoral district, and as long as it had more than four percent of the national vote, it is eligible for a compensatory seat (Aardal 2011). Both provincial seats and compensatory seats are apportioned using the

Figure 7.2 Electoral disproportionality over time in Norway
Source: Gallagher's Index (1991).

modified Sainte-Laguë method.[11] The Sainte-Laguë method reduces the bonus for large parties and therefore produces more proportional outcomes than other electoral formulas, such as d'Hondt (Aardal, 2011: 6).

Despite the introduction of the Sainte Laguë method in 1952 and compensatory seats in 1989, parties' national vote shares do not correspond perfectly with seat shares. In other words, some disproportionality exists in the Norwegian electoral system, as illustrated in Figure 7.2, which reports Gallagher's disproportionality index for each of Norway's legislative elections since World War II. Gallagher's index measures the difference between the percentage of *national* votes received by a party, and the percentage of seats a party receives in the resulting legislature. Deviations from proportionality decreased significantly after World War II, and particularly since 1989 when compensatory seats were introduced (Sørensen 2003, Aardal 2011).

Disproportionality emerges because most of Norway's legislative seats are awarded to parties by district in proportion to their share of district

[11] In November 1952, the electoral system was changed from the d'Hondt to the Sainte Laguë method for calculating the distribution of seats. In the subsequent 1953 election, the Labor Party lost six seats as a consequence of the shift from d'Hondt to Sainte Laguë (Aardal 2011).

votes (Matthews and Valen 1999, Sørensen 2003, Aardal 2011).[12] In other words, disproportionality is a result of having multiple subnational districts in a PR systems. In 2009, for example, the Labour Party would have won sixty seats if the country was organized as one, nationwide electoral district (Aardal 2011). But based on the district-level distribution of seats, the party won sixty-four (Aardal 2011).[13] While the Labour Party was better off thanks to Norway's multiple districts, some parties were made worse off. The Senior Citizen Party, for example, would have won a seat in the legislature if the entire country had been one electoral district in 2009. Yet, it received no provincial seats and was not eligible for a compensatory seat because it did not reach the national threshold of 4 percent. In sum, parties' national vote share does not correspond perfectly with their legislative seat share because most legislative seats are allocated according to parties' *district* vote shares rather than their national vote shares.

One other important characteristic of Norway's electoral system is the de facto closed party lists. In Norway, voters generally cast a ballot for a party list rather than individual candidates. The names on a party's list correspond with the candidates representing that particular party. These candidates are chosen by the nomination conventions of each party (Sørensen 2003). In theory, voters may modify the order of candidates on the list. Voters can change the rank order of the candidates on the party list and even cross out candidates if they so choose (Aardal, 2011: 8). However, the levels of coordination required to overturn the parties' rankings are so extreme that they deter most attempts to do so. At least half the voters have to make exactly the same alterations of the list for it to have any effect (Aardal, 2011: 8). For all practical purposes, Norway's system is effectively a closed-list system (Aardal, 2011: 8).

INCENTIVES TO TARGET

Given Norway's electoral institutions, what incentives, if any, do parties have to geographically target subsidies? Conventional wisdom suggests there will be little geographic targeting of economic benefits in a country like Norway. However, this widely held belief emerges from models of proportional representation that ignore geography (e.g. Grossman and Helpman 2005). In reality, the district-level allocation of seats that occurs

---

[12] And because some of Norway's districts have a relatively small number of seats (Aardal 2011, Carey and Hix 2011).
[13] Although the Labour Party kept all these seats, it was not eligible for any of the compensatory seats.

in most PR countries influences parties' election strategies and subsequently their policy priorities. Parties competing in proportional systems with multiple districts must be mindful not only of their national appeal but also of their support in each district. Focusing exclusively on maximizing the party's national vote share could cost the party a "provisional seat" (i.e. a seat allocated at the district level). This is particularly likely if the party's supporters are unevenly spread across the country's electoral districts. At the same time, however, ignoring a party's national vote share may make a party ineligible for a compensatory seat. A Norwegian political party made precisely this mistake in the 1989 election. The party, People's Action Future for Finnmark (*Folkeaksjonen Framtid for Finnmark*), focused exclusively on improving economic conditions in Finnmark, a district whose local economy had been badly hurt by poor fishing output, via state aid. The party won 21.5 percent of the vote in Finnmark and consequently received a "provincial" seat in parliament. However, the party was not eligible for a compensatory seat because it failed to clear the national threshold of 4 percent. The party won just 0.3 percent of the national vote in the 1989 election, which in unsurprising given the party's exclusive focus on regional issues.

In PR systems with multiple electoral districts, like Norway, the best electoral strategy is to win those votes that maximize the party's legislative seats. To achieve this goal, parties may seek to target benefits to select districts. Targeting will be especially useful for winning legislative seats if a party's supporters are unevenly distributed across electoral districts. If partisans are concentrated in some districts but not others, parties in PR systems will do well by targeting benefits to districts in which there are a large number of party supporters (Cox and McCubbins 1986, Levitt and Snyder 1997, Balla et al. 2002, Costa-i-Font, Rodriguez-Oreggia, and Lunapla 2003, Calvo and Murillo 2004, McGillivray 2004, Golden and Picci 2008). Cultivating areas of core support, where it is less expensive to attract the marginal supporter, is an efficient way to win additional legislative seats when seats are allocated to parties by districts in proportion to parties' district vote share.

Targeting assistance to party strongholds also helps to keep party supporters loyal. If a party withdrew aid from a party stronghold, it may lose voters to other parties and new parties might emerge to represent the disaffected voters (Golden and Picci 2008). The Labour Party's failure to provide sufficient economic support to the Norwegian district of Finnmark resulted in the emergence of a new party, the aforementioned People's Action Future for Finnmark (*Folkeaksjonen Framtid for Finnmark*). This new party emerged to demand increased government assistance to improve the economic conditions in Finnmark where unemployment had increased sharply due to shrinking fish resources in the district's coastal waters.[14] Although full employment was one of the fundamental goals of the Norwegian Labour Party since the 1930s (Aardal, 1990: 153), the party did not provide economic assistance to the region at levels believed to be sufficient by many voters. Labour's inability to target sufficient economic assistant to Finnmark was likely due to the coalition dynamics at the time. Labour failed to receive a majority mandate from voters in the 1985 election and governed as a minority government with the support of the right-wing Progressive Party. As a result, this period was one of the "most turbulent in the *Storting* since World War II" (Aardal, 1990: 152). Although Finnmark was traditionally a Labour stronghold, many voters felt that the Labour Party had not done enough to help the region. As a result, the new political party won one of the district's seats previously held by Labour in the 1989 election. Strikingly, it was the first post-war election in which Labour won fewer than two seats in Finnmark (Svåsand, Strøm, and Rasch, 1997: 96).

As the Finnmark example makes clear, targeting benefits to safe districts can help parties hold core voters and prevent the emergence of new parties (Golden and Picci 2008). In contrast, targeting assistance to districts with stiffer electoral competition entails greater risk and potentially fewer rewards, particularly in PR systems where multiple parties compete in multimember districts. In such systems, it is difficult to know precisely where the marginal seats are located (Sørensen, 2003: 171). Because political parties tend to be risk adverse, they focus their efforts on "safer" electoral strategies, such as targeting economic benefits to party strongholds (Cox and McCubbins 1986).

Parties can identify their core areas of support – even in multiparty PR systems with multiple districts. Their core areas of support are simply those districts where they win the largest share of the district's seats. The Conservative Party (*Høyre*), for example, won seven of Akershus' seventeen seats in the 2013 election. They did so by winning slightly more than 40 percent of the district's votes. Seeing this result, the Conservative Party knew that Akershus was a party stronghold. Targeting benefits to "safe" districts like Akershus entails fewer risks for the Conservative Party than trying to identify marginal seats in other districts. Because parties

---

[14] The party was formed by a man named Anders Aune who was the district's top public servant (*Fylkesmann*). For this reason, the party in sometimes referred to as the Aune list. It is also known as People's Action Future for Finnmark (*Folkeaksjonen Framtid for Finnmark*)

tend to be risk adverse, those competing in multidistrict PR systems will work to target benefits to "safer" districts (i.e. party strongholds) (Cox and McCubbins 1986), all else equal.

In closed-list systems, like Norway, parties can successfully target benefits to safe districts because they have firm control over individual party members. In open-list systems, targeting is more difficult because parties are less able to discipline their own members of parliament. Undisciplined legislators seek to target benefits to their core constituents who are typically localized in bailiwicks (Ames 1995). In Italy, for example, where open lists were used from 1953 to 1994, governing parties could not discipline their own members of parliament sufficiently to target the parties' areas of core electoral strength (Golden and Picci 2008). Instead, powerful individual legislators were able to secure resources for their constituents at the expense of the governing parties (Golden and Picci 2008). In contrast, closed-party lists engender sufficient discipline to allow parties to adeptly target benefits to their electoral strongholds.[15] Given this, I hypothesize that subsidies will flow disproportionality to safe districts in closed-list PR systems, like Norway. More precisely, I anticipate that subsidies will flow disproportionality to districts where the largest government party won by a greater margin over the next closest party in the previous election. I focus on the largest government party because it is best placed to target aid to its supporters in a multiparty coalition, particularly when it holds the relevant ministry.

POLICY TARGETING IN PRACTICE

Government parties have the ability to target economic benefits, such as subsidies, to select districts. In Norway, for example, government parties can target subsidies in at least two ways. First, the national government decides how much money to spend on subsidies for each sector of the economy. If sector employment is unevenly distributed within a country, the largest government party can target select districts via sector-specific

---

[15] Norwegian legislators are frequently lobbied by local interests and business organizations (Alt et al. 1999). Sixty-eight percent of legislators were contacted by business organizations on a weekly basis, and about the same amount had contacts with trade unions and professional groups (54 percent). Fifty-four percent of firms reported contacting a Member of Parliament at least once in the past year (Alt et al. 1999). Norwegian legislators report that lobbying activities have increased during recent years, and that lobbying increasingly influences spending decisions and government policy (Sørensen 2003).

subsidy budget allocations. Second, the government indirectly controls the allocation of subsidies to firms in a sector via the bureaucracy.

Sector Targeting

In Norway, the national government directly controls the funding of sector-specific subsidies. The government decides how much money to spend on subsidies for each sector of the economy. The amount of money allocated to a sector is renegotiated every year within the governments' budget process.[16] "In practice, last year's allocations often work as a starting point when allocations for the coming year are to be negotiated."[17]

Both political and economic considerations shape the government's funding decisions. In deciding how much money to allocate to a sector, there is "room for political priorities, for example if something unexpected happens and an industry crisis occurs."[18] Ultimately, the amount of money allocated to subsidies is determined by "political and strategic deliberations."[19]

Negotiations with sector-specific interest groups influence the government's funding decisions. For example, the main farmers' organizations (*Norges Bondelag* and *Norsk Bonde- og Småbrukarlag*) negotiate with the government every year over the agriculture-sector subsidy budget. Both the amount of money and the main guidelines for the expenditures are negotiated. In this way, interest groups enjoy a direct means of influence over governmental subsidy decisions.

Following negotiations with sector-specific interest groups, each ministry prepares its subsidy budget proposal. For example, the Ministry of Agriculture and Food prepares the budget for subsidies to the agriculture sector. The Ministry for Industry and Trade prepares the budget for subsidies to the manufacturing sector. The proposed budget is based on input from the various units of the ministry, including input from the underlying businesses and other relevant organizations, such as the aforementioned farmers' organizations and Innovation Norway, the main bureaucracy responsible for allocating subsidies. The individual

---

[16] Jardar Jensen, State Secretary to the Minister of Local Government and Modernisation, email communication, July 8, 2015.
[17] Bjørn Kåre Molvik, Deputy Director General of the subsidy section of the Ministry of Trade, Industry and Fisheries, email communication, August 17, 2015.
[18] Jardar Jensen, State Secretary to the Minister of Local Government and Modernisation, email communication, July 8, 2015.
[19] Jardar Jensen, State Secretary to the Minister of Local Government and Modernisation, email communication, July 8, 2015.

ministries' proposals are then put forward to the Ministry of Finance who prepares the final budget. The final subsidy budget is presented to Parliament by the Ministry of Finance for approval.

Although Parliament must approve the final budget, individual legislators and opposition parties have little direct influence over subsidies. Parliament "typically does not change the amount of money that has been agreed by the government and interest groups."[20] Instead, the government's allocation decisions are normally approved with no amendments or modifications. Given this, government parties enjoy relative autonomy over the allocation of money to economic sectors.

The government does not include money for specific firms or companies in the annual budget.[21] Instead, decisions regarding firm-level subsidies are made by civil servants. In this way, bureaucrats are the last link in the parliamentary chain of delegation (Strøm 2000). The final link in the chain of delegation is tightly controlled; the government exerts rigorous oversight of these bureaucrats and their subsidy decisions, as described in the following section.

Firm Targeting

Formally, bureaucrats decide which firms to subsidize within a sector using the monies allocated to the sector by the government.[22] In other words, bureaucrats have autonomy over firm-level subsidy allocation decisions. However, they are accountable to cabinet-level ministers (Rodrik 2004). Close monitoring (and coordination) of subsidy activities by a cabinet-level politician – that is, a "principal" who has internalized the optimal reelection strategy for her party – is necessary for subsidies to be an effective vote-winning policy tool.

Government ministers purposefully attempt to influence bureaucratic behavior (McCubbins, Noll, and Weingast 1987). To control bureaucratic decisions over subsidies, the government uses several mechanisms including budgets, letters of assignment, and biannual meetings. These mechanisms of control exist because of rational choices by politicians who care about the outcomes from bureaucratic behavior (Huber, Shipan, and Pfahler 2001). Bureaucratic behavior regarding subsidies is especially

---

[20] Siri Lothe, Senior advisor, Department of Agriculture, Ministry of Agriculture and Food, email communication on July 2, 2015.
[21] In person interview at Innovation Norway in Oslo, Norway with Sigrid Gåseidnes on June 23, 2014.
[22] In practice, the distinction between sector subsides and firm subsidies may be less clear if, for example, a single firm dominates a sector.

important to politicians because subsidies can help political parties win votes and subsequently seats (Buts et al. 2012).

In Norway, one of the principal bureaucracies charged with the allocation of subsidies is Innovation Norway (*Innovasjon Norge*).[23] Historically, Innovation Norway's mandate was limited to nonagriculture sectors but in recent years, it has become responsible for the agriculture sector as well.[24] Innovation Norway is, in theory, responsible for allocating subsidies to firms within a given sector. However, the government uses various mechanisms to indirectly control the allocation of firm-level subsidies, including, for example, the national budget.[25]

Budgets have long been recognized as a mechanism by which ministers and legislators can influence civil servants (Niskanen 1971, Banks 1989, Dunleavy 1991, Huber 2000). In Norway, the government uses the national budget to control the allocation of subsidies by specifying the total amount of money available for subsidies to specific sectors of the economy, such as manufacturing. Upon approval of the budget by Parliament, the government says to Innovation Norway, "here is the total budget for manufacturing. This money can go only to firms in the manufacturing sector."[26] The government gives each sector a "budget code."[27] Innovation Norway then charges all sector-specific subsidy programs to the appropriate budget code.[28]

Bureaucrats cannot spend more on subsidies for a given sector than is stipulated in the government's budget. Neither can bureaucrats reallocate funds from one sector to another. Bureaucrats may want to spend more money on manufacturing subsidies and less money on agriculture subsidies, for example.[29] The Norwegian agriculture sector is geographically diffuse

---

[23] The other bureaucracy charged with subsidy allocation is the Research Council of Norway.
[24] In person interview with Innovation Norway staff members Pål Aslak Hungnes and Per Melchior Koch in Oslo, Norway on June 19, 2014. See also Innovation Norway (2014b).
[25] Previous studies of bureaucracies have suggested several possible strategies for control, including the use of the budget processes (e.g. Bendor, Taylor, and Van Gaalen 1987, Banks 1989,) and ongoing oversight (e.g. Aberbach 1990). However, most research focuses on statutory control, whereby legislators use legislation to influence agency decision (e.g. Huber, Shipan, and Pfahler 2001).
[26] In person interview at Innovation Norway in Oslo, Norway with Sigrid Gåseidnes on 23 June, 2014.
[27] In person interview at Innovation Norway in Oslo, Norway with Sigrid Gåseidnes on 23 June, 2014.
[28] In person interview at Innovation Norway in Oslo, Norway with Sigrid Gåseidnes on 23 June, 2014.
[29] In person interview with Innovation Norway staff members Pål Aslak Hungnes and Per Melchior Koch in Oslo, Norway on June 19, 2014.

and politically powerful[30] and as a result, it wins generous government subsidies in Norway's closed-list PR system. Although unelected bureaucrats may view lavish agriculture subsidies as being economically inefficient, they cannot unpick this political outcome. Bureaucrats are constrained by the governments budgeting procedures; they cannot reallocate funds from agricultural subsidies to manufacturing subsidies because the government allocates subsidy funding by sector.

Bureaucrats would prefer to receive money from the government with "no strings attached."[31] A single subsidy budget without sector-specific allocations would give bureaucrats more autonomy to decide how to allocate subsidies.[32] With an "untied budget" from the government, bureaucrats could allocate money to sectors as they see fit. Yet, the government chooses not to give bureaucrats this level of autonomy. Politicians "don't want to lose control of subsidies" because they are a useful electoral tool.[33] Instead of giving Innovation Norway a big pot of money with no strings attached,[34] the government instead says, "these moneys are for agriculture" and asks Innovation Norway to allocate the designed funds to agriculture producers.[35]

Several mechanisms give the government indirect control over which producers receive subsidies within a given sector. High-level, semiannual meetings provide ministers with an opportunity to influence bureaucrats' decisions.[36] Twice a year, staff from Innovation Norway meet with Cabinet Ministers and their senior staff to discuss the allocation of subsidies. As the State Secretary to the Ministry for Local Government and Modernisation said, "the meetings provide a platform to discuss the annual reports, the finances and to develop a shared vision for the year to come."[37] These twice-yearly meetings provide the government with an opportunity to exert control over the bureaucracy and their decisions.

---

[30] See Chapter 6.
[31] In person interview with Innovation Norway staff members Pål Aslak Hungnes and Per Melchior Koch in Oslo, Norway on June 19, 2014.
[32] In person interview with Innovation Norway staff members Pål Aslak Hungnes and Per Melchior Koch in Oslo, Norway on June 19, 2014.
[33] In person interview with Innovation Norway staff members Pål Aslak Hungnes and Per Melchior Koch in Oslo, Norway on June 19, 2014.
[34] In person interview at Innovation Norway in Oslo, Norway with Sigrid Gåseidnes on June 23, 2014.
[35] In person interview at Innovation Norway in Oslo, Norway with Sigrid Gåseidnes on June 23, 2014.
[36] Legislators are not involved in these processes. (Jardar Jensen, State Secretary to the Minister of Local Government and Modernisation, email communication, July 8, 2015).
[37] Jardar Jensen, State Secretary to the Minister of Local Government and Modernisation, email communication, July 8, 2015.

Annual letters of assignment also provide the government with a mechanism of control over bureaucratic actions. Bureaucrats charged with dispersing subsidies receive annual letters of assignment from the relevant ministry. Innovation Norway, for example, receives yearly assignment letters from the Ministry of Trade, Industry and Fisheries, the Ministry of Local Government and Modernisation, the Ministry of Agriculture and Food, and the Ministry of Foreign Affairs. Based on the national budget, the letters of assignment set out spending limits that stipulate the amount available for new loans and subsidies for a given sector of the economy (Innovation Norway 2014b). The letters also stipulate strategic and operational guidelines related to subsidies.[38]

The more detailed these letters are, the stronger the constraints they impose on bureaucratic behavior (Huber, 2000: 400, Huber and Shipan 2002). Detailed letters may include, for example, precise instructions regarding the allocation of subsidies to firms within a given sector. A detailed letter may also specify the government's explicit expectations and requirements of the bureaucrats' activities and decisions.[39] In contrast, a letter that stipulates only the annual budget for a sector leaves more room for bureaucratic discretion.

Government ministers appear to understand the constraints imposed by detailed letters of assignment.[40] The State Secretary to the Minister of Local Government and Modernisation identified "the number of details in these letters" as a key mechanism by which the Ministry sought to limit Innovation Norway's discretion in subsidy allocation decisions.[41] Some ministries, including the Ministry of Trade, Industry and Fisheries, admit to using the letters of assignment to stipulate precisely what areas should be prioritized.[42] In short, annual letters of assignment provide a means for ministerial control over civil servants.

Government parties seek to control the allocation of subsidies because subsidies win votes. Given that subsidies are a vote-winning policy tool, why would governments ever delegate subsidy decisions to unelected bureaucrats in the first place? Surely, governments would want to

---

[38] Jardar Jensen, State Secretary to the Minister of Local Government and Modernisation, email communication, July 8, 2015.
[39] Jardar Jensen, State Secretary to the Minister of Local Government and Modernisation, email communication, July 8, 2015.
[40] Similarly, legislation with a vague – as opposed to a specific policy mandate – allows bureaucrats relatively more autonomy (Epstein and O'Halloran 1994, Huber and Shipan 2000, Huber, Shipan, and Pfahler 2001).
[41] Jardar Jensen, State Secretary to the Minister of Local Government and Modernisation, email communication, July 8, 2015.
[42] Bjørn Kåre Molvik, Deputy Director General of the subsidy section of the Ministry of Trade, Industry and Fisheries, email communication, August 17, 2015.

control the allocation of subsidies themselves to maximize their electoral rewards? But by controlling the allocation of firm-level subsidies only indirectly, the government insulates themselves from rent seeking. For example, firms sometimes approach ministers directly to request a subsidy.[43] But ministers defer such requests to Innovation Norway.[44] In this way, the bureaucracy shields ministers from rent-seeking. This institutional design may be a purposeful effort to minimize rent-seeking (Rodrik 2004).[45] Indeed, ministers report that they appreciate being able to pass on subsidy requests to Innovation Norway.[46] Doing so gives them "political cover" if the government is unable or unwilling to satisfy the request. By delegating subsidy decisions to unelected bureaucrats, governments have the best of both worlds: they can exert control over the allocations of subsidies for electoral gain and at the same time they can "scapegoat" bureaucrats for unpopular decisions (Remmer 1986, Vreeland 2003). This type of delegation provides maximum electoral benefits.

EMPIRICAL TESTS

I argue that the distribution of subsidies over all electoral districts within a country will exhibit a political bias. More precisely, I hypothesize that subsidy spending per manufacturing sector employee will be relatively higher in districts where the largest government party won by a greater margin in the previous election. To test this proposition, I regress government spending on manufacturing-sector subsidies per employee in each of Norway's electoral districts on a measure of electoral competitiveness.

Measuring Electoral Competitiveness

The most commonly employed measure of electoral competitiveness, whether at the district level (e.g. Mayhew 1974, Aidt, Golden, and Tiwari, 2011) or cross-nationally (e.g. Anderson and Beramendi 2012) is the difference in vote share between the top two parties.[47] This measure is often called the vote margin. Within-country vote margins calculated at

---

[43] In person interview with Innovation Norway staff members Pål Aslak Hungnes and Per Melchior Koch in Oslo, Norway on June 19, 2014.
[44] In person interview with Innovation Norway staff members Pål Aslak Hungnes and Per Melchior Koch in Oslo, Norway on June 19, 2014.
[45] In person interview with Innovation Norway staff members Pål Aslak Hungnes and Per Melchior Koch in Oslo, Norway on June 19, 2014.
[46] In person interview with Innovation Norway staff members Pål Aslak Hungnes and Per Melchior Koch in Oslo, Norway on June 19, 2014.
[47] See Strøm (1992) for a theoretical definition of competitiveness.

the district level are informative measures of electoral competitiveness, even in multiparty PR systems (Kayser and Lindstädt 2015).

In a multiparty government coalition, the largest party is best placed to strategically allocate subsidies for electoral gain, particularly when it controls the relevant ministry. Therefore, I calculate the largest government party's vote margin in Norway's nineteen electoral districts for two regularly scheduled, national legislative elections in 2005 and 2009. I calculate the difference between the largest government party's vote share and the next closest party in each electoral district.

In parliamentary systems like Norway, the cabinet rather than the legislature constitutes the government. The cabinet consists of a portfolio of departments or ministries, such as the Ministry of Finance. Because no party won an absolute majority of legislative seats in the 2005 election, the cabinet included three parties: the Labour Party (DNA), the Centre Party (*Senterpartiet*), and the Socialist Left Party (*Sosialistisk Venstreparti*). The 2005 cabinet was the first time the Socialist Left Party sat in government. This minimum winning cabinet was known as the "Red–Green coalition." The Labour Party, which held the largest share of parliamentary seats (36 percent) after the 2005 election, also held the largest share of cabinet seats. The Labour Party held ten cabinet seats; the Socialist Left Party had five cabinet seats and the Centre Party had four.[48]

As the largest party in government, the Labour Party secured both the prime ministership and the Ministry of Trade and Industry, which oversees manufacturing-sector subsidies. In a multiparty coalition, each party is generally able to implement its own priorities in the areas under its ministries' jurisdiction (Laver and Shepsle 1994). By holding the Ministry for Trade and Industry, the Labour Party was uniquely well-placed to direct manufacturing subsidies to districts where they did especially well in the 2005 election.

The three-party Red-Green coalition won reelection in 2009. The Labour Party retained both the prime ministership and the Ministry of Trade and Industry. Effectively, the Red-Green government continued in office with little change after the 2009 election. Despite being the largest party in the Red-Green coalition in 2005 and 2009, Labour's vote margin varied between electoral districts. In 2005, for example, Labour's largest vote margin was in the district of Hedmark where it won 45.89 percent of the votes cast – 28.88 percentage points more than the next largest party. Given this convincing win, Hedmark can be characterized as a "safe"

---

[48] The opposition consisted of four parties: the Progress Party, the Conservative Party, the Christian Democratic Party, and the Liberal Party.

district for Labour. Labour did not fare equally well in all districts. In Vest-Agder, for example, Labour faced tough competition from the Progress party. In 2005, the Progress party won 23.95 percent of the vote in Vest-Agder while the Labour Party won 23.93 percent. In this highly competitive district, just 0.02 percentage points separated the two parties' vote share.

The variable, *Vote Margin,* equals the difference between the largest government party's vote share and the next closest party in the most recent previous election. For example, the variable *Vote Margin* equals −0.02 for the district Vest-Agder because Labour, the largest government party, was 0.02 percentage points behind the next closest party (Progress) in this district in the 2005 election.

THE EMPIRICAL MODEL

I regress government spending on manufacturing-sector subsidies per employee in each of Norway's nineteen electoral districts on *Vote Margin*. By calculating subsidies per employee, I effectively control for economic geography. Because of the uneven geographic distribution of economic activities, each electoral district contains different numbers of manufacturing employees. The geographic distribution of manufacturing-sector employees is what I refer to as "economic geography." Economic geography, together with electoral institutions, helps to explain the variation in subsidies between countries and within countries between sectors (see Chapter 6).

To control for economic geography, I calculate the amount of money spent on sector-specific subsidies per sector employee in each district. This measure accounts for the uneven distribution of manufacturing employees between electoral districts. It provides a measure of subsidy spending that is comparable between districts. In this way, I can isolate the potential effects of district-level electoral competitiveness.

I regress government spending on manufacturing-sector subsidies per employee in each of Norway's electoral districts on *Vote Margin*, holding several factors constant. First, I control for districts' unemployment rate because districts with relatively higher rates of unemployment may receive more generous subsidies from the government. Governments may seek to encourage employers to hire new workers using subsidies and districts with higher unemployment rates may therefore receive more government assistance. The district Rogaland, for example, received the lowest subsidy amount per employee – just 1,065 Norwegian krone (NOK) on average over the period from 2005 to 2012. The district's low unemployment rate may explain why it received so little government assistance. I therefore include the unemployment rate as a control variable in all estimated models.[49]

I also control for the population density of each district. The Norwegian government has a long history of working "to spread business across the country by subsidizing producers in rural areas."[50] As the State Secretary to the Minister of Local Government and Modernisation said, "The main objective (of subsidies) is to achieve value creation and economic growth in all regions of Norway."[51] The government will, for example, fund a building in a rural area that costs more than it is worth because it is in an isolated area with no secondary market/capital value.[52] Because of this strategy, rural districts with lower population density may receive relatively more subsidies, all else equal. Population density therefore serves as an important control variable.[53]

Labour's district-level vote margins do not correspond closely with population density. Among the more densely populated southern districts, Labour wins by varied amounts. For example, Labour won by large margins in some southern districts, such as Hedmark and Oppland, but obtained much smaller margins in others. Similarly, Labour's vote margins vary among the less densely populated northern districts. Labour won by large margins in Finnmark but faced much stiffer competition in Troms – despite the fact that both districts are sparsely populated. In sum, the cross-district variation in support for the Labour Party does not correspond closely with population density or the country's north-south divide.[54]

---

[49] The unemployment variable equals the number of unemployed persons in a district as a percentage of the district's population. Unemployment captures the economic performance of a district. Alternative measures of a district's economic performance might include GDP, GDP per capita and/or poverty rates. Unfortunately, these data are unavailable for much of the sample period. For example, GDP by district is available only from 2011 and household poverty measures are only available from 2013. When both measures are available, GDP and unemployment are highly negatively correlated (−0.94). The correlation between GDP and subsidies is negative but modest (−0.4). Oslo is the richest county and yet it falls within the second quartile in terms of subsidies. In fact, Oslo receives nearly the same about of subsidies per person as the second poorest county: Aust-Agder.

[50] In person interview with Innovation Norway staff members Pål Aslak Hungnes and Per Melchior Koch in Oslo, Norway on June 19, 2014.

[51] Jardar Jensen, State Secretary to the Minister of Local Government and Modernisation, email communication, July 8, 2015.

[52] In person interview with Innovation Norway staff members Pål Aslak Hungnes and Per Melchior Koch in Oslo, Norway on June 19, 2014.

[53] This logic is similar to the logic underlying the measure of "relative" geographic concentration used in previous chapters, which captures the degree of a sector's employment concentration relative to the geographic distribution of total employment.

[54] Similarly, in Sweden, district-level population density does not correlate with parties' vote share (Rodden, unpublished manuscript). However, Figure 7.1 suggests

Voter turnout may also influence the distribution of subsidies. Government parties may target areas that provide the best return in terms of votes (Martin 2003). As a result, electoral districts with higher turnout may receive relatively more subsidies, all else equal. I therefore control for each districts' turnout rate in the previous national election.

The size of a district's legislative delegation may also be important in securing resources for the district (Ansolabehere, Gerber, and Snyder 2002). Therefore, I also control for district magnitude. In Norway, district magnitude ranges from 4 to 19. Since 2005, the number of legislators per district is a function of the district's area and population (Aardal 2011).

I estimate ordinary least-squares regressions with robust standard errors and year-fixed effects. The inclusion of year-fixed effects ensures that any national-level shocks, such as year-to-year fluctuations in oil prices or economic crises, are absorbed by the year-fixed effects. In 2009, for example, the government significantly increased the total subsidy budget as part of a nationwide economic crisis package.[55] Year-fixed effects control for omitted variables that vary over time but are constant across districts. Year fixed effects also ensure that the focus on is the cross-district variation in subsidies, which is precisely the variation in which I am interested. However, including year-fixed effects sets up a conservative test of the hypothesis.[56] The unit of analysis is district-year and my sample includes all of Norway's electoral districts during the period from 2006 to 2012.

---

that subsidies in Norway tend to be more generous in districts in the north of the country, as compared to the south. The districts with the two highest subsidy amounts per employee, Troms and Finnmark, are both in the far north of the country. The entire district of Troms is north of the Arctic Circle, and Finnmark is located at the very top of Norway adjacent to Russia and Finland. In contrast, Rogaland and Vestfold are located in the south. Rogaland is the center of the Norwegian petroleum industry and as a result it is a relatively prosperous district. Vestfold is on the western side of the Oslo fjord and serves as a commuter belt for the capital city. Vestfold is also home to shipping and related industries as well as food-processing companies. Given this pattern, population density is an important control variable because it serves as a proxy for Northern districts, which are less populous than Southern districts.

[55] Sigrid Gåseidnes, Innovation Norway staff, email Communication, June 24, 2015

[56] All reported results are also robust to the inclusion of a lagged dependent variable. The amount of money allocated to sector-specific subsidies is renegotiated each year within the governments' budget process. However, "last year's [subsidy] allocations often work as a starting point when allocations for the coming year are to be negotiated" (Bjørn Kåre Molvik, Deputy Director General of the subsidy section of the Ministry of Trade, Industry and Fisheries, email communication, August 17, 2015).

RESULTS

Electoral competition influences the distribution of subsidies across districts. Safe districts win relatively more subsidies than swing districts in this closed-list PR system.[57] Manufacturing subsidies per employee are relatively more generous in districts where the largest government party won by a greater margin in the previous election. This result suggests that government parties try to consolidate the partisan advantage that helped them win office in the first place by targeting subsidies to loyal partisans in less competitive districts.

The positive relationship between vote margin and subsidies is illustrated by Figure 7.3, which plots the prediction for *Subsidy Amount* from a linear regression of the two-year lag of logged *Subsidy Amount* on *Vote Share*, along with a confidence interval. The actual observations are then overlaid and labeled with the name of the electoral district.

Districts where the Labour Party won a larger vote share in 2005 received more generous subsidies per employee in 2007.[58] For example, Finnmark, where the Labour Party received 22.6 percentage points more of the vote share than the next largest party, received the second highest subsidy amount per employee. In contrast, Vestfold received the lowest subsidy amount and the Labour Party's vote margin was less than one percentage point (0.93). The pattern illustrated by Figure 7.3 suggests that the Labour Party rewarded partisan strongholds (i.e. safe districts) with more generous subsidies.

Table 7.1 reports the results from the fully specified models. The estimated coefficient on *Vote Margin* is positive and statistically significant in all estimated models. It is also substantively large.[59]

---

[57] Of course, it is possible that a record of subsidies creates safe seats. Party lists that successfully bring subsidies to the district may win more votes and thus engender safer districts. It is difficult to tease out which comes first: subsidies or votes. However, in my empirical tests, I treat votes as primary and regress vote share on subsidies. Votes in the previous election correlate with subsidies in subsequent years. It is also worth noting that senior legislators are not "parachuted" into safe district in Norway, as they are in other countries, such as France. In Norway, candidate selection procedures are highly decentralized (Matthews and Valen, 1999: Chapter 4). Local party officials select the candidate for the party's district list (Matthews and Valen, 1999: Chapter 4). Parachuting in a nonlocal candidate is unlikely to be a successful electoral strategy (Kaare Strøm, personal communication, May 4, 2016). In fact, voters would probability punish such attempts (Kaare Strøm, personal communication, May 4, 2016).

[58] I use 2007 spending data for this illustrative example to ensure that the new government coalition has sufficient time to influence subsidy spending. Using the 2006 spending data produces a similar graph.

[59] In Table 1, subsidies per employee are logged and as a result the coefficients are difficult to interpret directly.

Table 7.1 *Explaining the variation in manufacturing subsidies per employee between electoral districts*

	(1)	(2)	(3)	(4)	(5)	(6)
L.Vote Margin	0.047***	0.048***	0.048***	0.041***	0.026***	0.026***
	(0.008)	(0.007)	(0.007)	(0.007)	(0.008)	(0.007)
L.Population Density			-0.001***	-0.001***	0.0001	0.0001
			(0.000)	(0.000)	(0.000)	(0.000)
L.Unemployed(%)				76.56***	-37.56	-32.10
				(26.14)	(31.85)	(32.33)
L.Turnout(%)					-0.260***	-0.227***
					(0.049)	(0.062)
L.District magnitude						-0.029
						(0.025)
Constant	7.449***	7.268***	7.321***	6.037***	28.160***	25.713***
	(0.139)	(0.246)	(0.251)	(0.531)	(4.258)	(5.101)
Year fixed effects	No	Yes	Yes	Yes	Yes	Yes
Observations	133	133	133	133	133	133
R-squared	0.174	0.352	0.370	0.403	0.529	0.533

Notes: Robust standard errors in parentheses; *** $p < 0.01$, ** $p < 0.05$, * $p < 0.1$. Data cover Norway's nineteen electoral districts from 2006 to 2012.

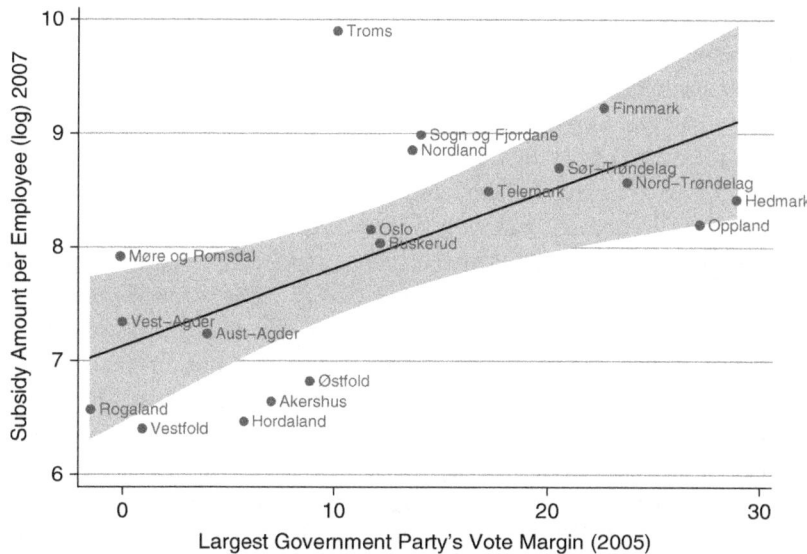

Figure 7.3 Largest government party's vote margin and subsidies per employee, by district
Source: Author's calculation using subsidy data provided by Innovation Norway and election returns from Statistics Norway.

Increasing *Vote Margin* from 4 points (i.e. Labour's vote margin in Aust-Agdar) to 29 points (i.e. Labour's vote margin in Hedmark) increases subsidies by NOK 3,450 ($415) per manufacturing sector employee in the most conservative model.[60] In sum, subsidies flow disproportionality to "safe" districts in this closed-list PR systems, all else equal.[61]

CONTROL VARIABLES

More densely populated districts receive fewer subsidies per employee, all else equal. In other words, subsidies flow disproportionally to rural districts with low population density. This result is consistent with the government's aspiration to spread business more evenly across the

---

[60] Including both year fixed effects and a lagged dependent variable reduces the magnitude of the coefficient on *Vote Share*. The one-year lag of subsidy spending is highly significant and indicates that subsidy spending, like most types of government spending, is sticky and changes slowly over time. However, the coefficient on *Vote Margin* remains positive, statistically significant and substantively large in models that include a lagged dependent variable.

[61] This result is consistent with Naoi's (2009) finding that subsidies decline in the face of higher political competition.

country.⁶² However, the negative coefficient on *Population Density* loses statistical significance in models that include *Turnout*. It is important to note that both variables have population as their denominator and are positively correlated with one another (r = 0.3). However, *Vote Margin* remains robust to the inclusion of *Turnout*.

Turnout is negatively correlated with manufacturing subsidies. Districts with higher turnout rates receive fewer subsidies per manufacturing employee. However, it is important to note that voter turnout is generally quite high in Norway. The sample average is 76 percent with a standard deviation of two. The lowest rate of turnout is still more than 70 percent (i.e. 70.4 percent in Finnmark in 2005). Given the high rate of turnout for all districts, parties may eschew attempts to "turnout" additional voters and focus instead on rewarding party loyalists in safe districts.⁶³

In two out of three models, *Unemployment* does not robustly predict subsidy spending.⁶⁴ This null result may be due to multicollinearity. In models without *Turnout*, *Unemployment* is positive signed, as expected. Districts with higher unemployment receive more generous subsidies per person than districts with less unemployed persons. Yet, once *Turnout* is included the coefficient on *Unemployment* becomes insignificant.⁶⁵

District Magnitude is not a robust predictor of subsidies. Districts with more representatives in parliament receive no more generous subsidies than districts with fewer representatives. *Vote Margin* remains a robust predictor of subsidies even after controlling for district magnitude. District magnitude is negatively correlated with vote margin (r = -0.27). In other words, Labour wins less of the vote share in districts with more seats. Given this, one concern might be that district magnitude influences subsidy spending rather than vote share per se. However, vote margin is

---

⁶² In person interview with Innovation Norway staff members Pål Aslak Hungnes and Per Melchior Koch in Oslo, Norway on June 19, 2014.

⁶³ Alternatively, the negative coefficient on *Turnout* may be an artefact of multicollinearity between the explanatory variables. For example, unemployment and turnout are negatively correlated at -0.35. This correlation may explain why the introduction of *Turnout* changes the estimated coefficient on *Unemployment*. Regardless, the estimated coefficient on *Vote Margin* remains positive and statistically significant for all estimated models.

⁶⁴ Perhaps welfare spending flows disproportionality to districts with higher unemployment rates thereby "squeezing out" subsidies.

⁶⁵ In these models, electoral tactics appear to dominate economic concerns. Subsidies are allocated primarily according to the political characteristics of a constituency (i.e. competitiveness) rather than economic need. Mehiriz and Marceau (2013) come to a similar conclusion regarding grant allocation decisions in Quebec, Canada.

robust to the inclusion of district magnitude and district magnitude never reaches conventional levels of statistical significance. These results suggest that it is the vote margin of the largest government party that matters for subsidy allocation rather than district magnitude. Presumably district magnitude does not matter for subsidy spending in Norway because it has no influence on politicians' election strategies in closed-list PR systems, as demonstrated in Chapter 6. Politicians have few incentive to cultivate a personal vote in closed-list PR systems, like Norway, and increases in district magnitude do nothing to change this. No matter how many seats are to be filled in a district, politicians in closed-list systems seek to appease party leaders rather than cultivate a personal vote.

CONCLUSION

In this chapter, I investigate the variation in government spending per employee on manufacturing subsidies between electoral districts in a closed-list PR country. Two novel results emerge. First, government parties competing in a country with closed party lists, proportional electoral rules, and multiple electoral districts, engage in electorally motivated policy targeting. This finding is unexpected; few would anticipate policy targeting in a country with electoral institutions like Norway's. Yet, the distribution of subsidy spending across electoral districts in Norway reveals evidence of electorally motivated policy targeting. Second, in this closed-list PR system, government parties target benefits disproportionality to electoral districts where they have relatively more supporters. Per employee, manufacturing subsidies are relatively more generous in districts where the largest party in government won by a greater margin in the last election, all else equal. In other words, government parties in closed-list PR systems target benefits to "safe" districts – that is, electoral districts with relatively large numbers of party loyalists.

Both findings run counter to conventional wisdom regarding policy targeting, which is derived largely from studies of plurality countries and the United States in particular (Golden and Min 2013). Research on this topic in plurality countries is dominated by a debate over whether parties target benefits to competitive (i.e. "swing") districts or safe districts. The evidence generally suggests that benefits flow disproportionality to swing districts or "competitive constituencies" in plurality systems (Golden and Min 2013). However, as I show here, the same is not true in closed-list PR systems.

In one of the first empirical studies of policy targeting in a closed-list PR country, I find evidence that the largest party in government disproportionality targets subsidies to the party's safe districts – that is, those district where they won a larger share of the vote in the last election. In Norway, an increase in the largest government party's vote margin of 25 percentage points correlates with an increase in subsidies to the district equal to NOK 3,450 ($415) per employee. Chapter 5 reported evidence of similar policy targeting in Austria, which, like Norway, has de facto closed party lists. Government parties in Austria supported a subsidy that disproportionality benefited areas where they had strong voter support, as discussed in Chapter 5.[66] The results from both Austria and Norway suggest that government parties in closed-list PR systems target particularistic economic benefits to safe districts, all else equal.

Policy targeting occurs even in the absence of personal vote seeking. Even in countries where politicians have little incentive to cultivate their own personal bases of support, such as Norway and Austria, policy targeting happens. This novel finding suggests that personal vote seeking is not a necessary condition for policy targeting. Instead, policy targeting can occur even in the absence of personal vote seeking. In Norway, for example, policy targeting happens when it helps parties maximize the number of legislative seats they control. In other words, personal vote seeking is not the only reason for geographically targeted economic policies.

---

[66] Intriguingly, the Austrian example suggests that district-level electoral competitiveness may have different effects on different parties in PR systems. Government parties may target benefits to safe districts, while opposition parties – unable to influence government spending – may focus their efforts on wining additional votes in competitive districts via other means. See Chapter 5 for further details.

8

Conclusion and Implications

This book examines why elected leaders in some democracies are more responsive to special interests than other. Leaders' responsiveness to interest groups depends on a country's electoral institutions and economic geography. Electoral institutions generate (re)election incentives for politicians and political parties and economic geography determines the optimal policy with which to respond to these incentives. As a result, economic geography mediates the effect of electoral institutions on policy outcomes.

Although economic geography plays an important role in policy making, many previous studies overlook the geographic distribution of economic activity. Purely institutional arguments posit a direct relationship between electoral institutions and policy outcomes. Because of the ubiquity of such arguments, it has become almost trivial to say that "institutions matter." Institutions certainly matter; however, their effects on economic policy are indirect. Electoral institutions generate (re)election incentives but economic geography determines the optimal policy with which to respond to these incentives. As a result, identical electoral institutions can produce different policy outcomes depending on patterns of economic geography.

In this book, I develop and test my argument in the context of economic policy. Economic policies that selectively assist small groups of citizens at the expense of many, such as government-funded subsidies, are my primary focus. Subsidies help people employed in a subsidized industry by raising wages and safeguarding stable employment, but they do so at the expense of taxpayers who ultimately fund such programs. In this way, subsidies redistribute wealth between groups of citizens in a country. Governments' decisions about subsidies reveal their political priorities. Governments that allocate more of their budgets to industrial subsidies are more willing to privilege small, narrow groups of voters over larger groups.

Money spent on industrial subsidies is money no longer available for health care or education. However, some governments instead prioritize programs that benefit larger groups of citizens at the expense of subsidies. The Swedish Prime Minister, for example, refused to fund subsidies for the automotive industry stating that he would not put "taxpayer money intended for healthcare or education into car companies" (Ward 2009). In this case, the Prime Minister explicitly prioritized broadly-beneficial programs like education over a narrowly-beneficial subsidy program. Yet, leaders in different countries make different policy decisions. The question is why. What explains the cross-national variation in government spending on subsidies? This variation is puzzling given existing institutional theories (e.g. Persson and Tabellini 2003). Why do governments in countries with identical electoral institutions exhibit varied spending priorities?

Governments elected via similar institutions spend dissimilar amounts on subsides because the usefulness of subsidies as an election-winning policy tool depends on economic geography. Economic geography refers to the geographic distribution of economic activity. The geographic distribution of economic activity determines the location of the beneficiaries of government-funded subsidies. If the beneficiaries are dispersed across many electoral districts, subsidies are an inefficient electoral tool in plurality systems where elections are won district-by-district. If the beneficiaries are geographically concentrated in a single district, a subsidy is roughly analogous to legislative particularism, or "pork barrel" spending. Bringing "pork" home to their own district helps politicians cultivate a personal base of support among voters in their district, which increases their reelection chances in plurality systems (Ferejohn 1974, Fenno 1978, Wilson 1986). As a result, geographically-concentrated groups tend to win more generous subsidies in countries with plurality electoral rules, all else equal.

Geographically diffuse industries win relatively greater subsidies in countries with proportional electoral systems. Subsidies for diffuse industries are politically expedient in PR systems where parties win legislative seats in proportion to their vote share. To maximize their vote share, parties supply subsidies to geographically diffuse groups. A subsidy to a geographically diffuse industry, for example, indirectly benefits more people than a subsidy to a concentrated industry. Subsidies help not only those people directly employed in an industry but also people working in related industries, such as retail and services. Subsidizing the geographically diffuse construction industry, for example, helps people employed in real estate, timber yards, and restaurants across the country. Because of these positive spillover effects, there is a "dispersion bonus" from subsidizing geographically-diffuse industries and the political profit from this dispersion bonus is greater in PR systems than in plurality systems.

EMPIRICAL CHAPTER OVERVIEW

In Chapter 4, I investigate the variation in government spending on manufacturing subsidies between democratic countries. Quantitative statistical tests confirm that electoral institutions and economic geography work together to explain the variance in manufacturing subsidies between democratic countries. Governments elected via plurality systems spend relatively more of their budget on subsidies when the beneficiaries are geographically concentrated. When subsidy recipients are geographically diffuse, governments in PR systems spend relatively more money to subsidize them, all else equal. These results are robust to alternative model specifications including those that relax the assumption that electoral systems are exogenous.

Chapter 5 provides qualitative evidence linking electoral institutions, economic geography, and economic policy. In this chapter, I examine two subsidy programs in countries with very different electoral institutions: Austria and France. In France, candidates win office by obtaining a majority of votes in the first ballot or failing that, a plurality of votes at the second ballot (Elgie 2005). Because of this electoral system, subsidies in France tend to go to geographically-concentrated groups (Verdier 1995),[1] as illustrated by the subsidy program examined in Chapter 5. Only producers located in a small, well-defined geographic area (i.e. approximately 1,000 hectares in the Cognac region) received financial assistance via this program. Legislators who represented the area, including those from different parties, pushed for the selective subsidy. Their demands were ultimately successful and the French government subsidized Cognac producers in violation of EU state aid rules.

While the French subsidy was highly selective and available only to producers in a concentrated geographic area, the Austrian subsidy, in contrast, was broadly beneficial and available to all farm-gate wine merchants regardless of their geographic location. Austrian farm-gate wine merchants, who benefited from the subsidy program, were spread

---

[1] Subsidies for geographically-concentrated groups are generously funded by the French government. In 2013, for example, subsidies equaled 0.6 percent of GDP in France.

across more than forty-five thousand hectares in Austria.[2] Subsidizing a geographically diffuse group like farm-gate wine merchants is politically expedient for parties competing in a country where elections are held via proportional electoral rules and closed-party lists. Because of the institutionally generated incentives that Austrian leaders faced, the government provided this subsidy to a geographically-diffuse producer group even though it violated EU state aid rules.

The French and Austrian examples demonstrate that the electoral incentives to subsidize producers sometimes outweigh the costs of violating international rules. The domestic incentives to supply subsidies are greatest when electoral institutions and economic geography align. Together, electoral institutions and economic geography predict the likelihood that an EU member-country will subsidize producers in violation of EU state aid rules, as demonstrated in Chapter 5. Illegal (i.e. noncompliant) subsidies are more likely in PR systems than plurality systems when the beneficiaries are geographically diffuse. Arguably this is because the electoral incentives to supply subsidies to diffuse groups are relatively larger in PR systems. When the potential electoral benefits are large, governments are willing to provide subsidies in violation of international economic agreements.

In Chapter 6, I examine government spending on subsidies in countries with proportional electoral systems. I examine the variation in subsidies both within and between PR countries. Looking at subsidies within a single PR country, I find that geographically diffuse sectors win more generous subsidies than geographically concentrated sectors, all else equal. This evidence demonstrates that the electoral support of geographically-diffuse groups is especially valuable in PR systems.

In all PR systems, the political bonus from subsidizing geographically diffuse groups is electorally valuable, and as a result, spending on subsidies is higher in PR systems when the beneficiaries are more diffuse. However, spending on geographically diffuse groups is relatively higher in closed-list PR systems, as compared to open-list PR systems. In open-list systems, some funds are diverted by powerful individual legislators to groups that are geographically concentrated in their own electoral district or bailiwick (Ames 1995, Golden and Picci 2008). Legislators in open-list PR systems have incentives to divert resources in this way to cultivate their own personal support base. Doing so helps them win more individual votes and consequently increases their reelection chances. As a result, spending on subsidies for geographically diffuse groups is relatively higher in closed-list PR systems where political parties can better

---

[2] Data from www.bmlfuw.gv.at.

discipline their legislators to ensure subsidies flow to geographically diffuse groups.

The difference between open-list and closed-list systems highlights an important mechanism linking electoral institutions, economic geography, and policy outcomes: the nature of electoral competition. Open-list systems engender candidate-centered elections because candidates must work to differentiate themselves from co-partisans in order to win preference votes. To do this, candidates highlight their own personal qualifications and work to (or promise to) provide select benefits to their district or bailiwick. The candidate-centered nature of elections in open-list systems, as in plurality systems, incentivizes the provision of economic benefits to geographically concentrated groups. These incentives grow stronger as district magnitude increases and politicians compete against even more co-partisans for voters' support. Among PR systems, the most generous subsidies occur in open-list systems with high mean district magnitude and geographically concentrated groups, as shown in Chapter 6. This finding demonstrates that the nature of electoral competition is an important mechanism linking electoral systems and economic geography to policy outcomes.

In Chapter 7, I investigate the variation in subsidies between electoral districts within an archetypal PR country: Norway. Two novel results emerge from this single-country study. First, political parties competing in closed-list proportional electoral systems engage in policy targeting – that is, they supply benefits to select, geographically defined groups. Second, political parties in this de facto closed-list PR system target economic benefits disproportionality to districts where they have relatively more supporters. Controlling for economic geography, districts where the largest government party won a greater share of the votes in the previous election receive more generous subsidies, all else equal. Government parties in this PR system use subsidies to consolidate the partisan advantage that helped them win seats in the first place by subsidizing producers in "safe" districts.

Chapter 7 also includes novel qualitative evidence from interviews of government ministers and bureaucrats responsible for subsidy programs. These interviews confirm the importance of electoral politics and economic geography for governments' spending decisions. The interviews also illustrate the mechanisms that governments use to target subsidies to politically important areas. In Norway, the government targets subsidies in at least two ways. First, the government decides how much money to spend on subsidies for each sector of the economy. Via the budget process, the government directly controls the allocation of subsidies to economic sectors, including agriculture and manufacturing. Second, the government indirectly controls the allocation of subsidies to firms within a given sector via control of the bureaucracy. To control bureaucratic decisions over firm-level subsidies, the government uses various mechanisms including letters of assignment and biannual meetings. These mechanisms of control exist because political parties care about the outcomes of bureaucratic behavior (Huber, Shipan, and Pfahler 2001). Bureaucratic decisions regarding subsidies are especially important because subsidies help political parties win votes and subsequently legislative seats (Buts et al. 2012).

CONTRIBUTIONS

This study makes several important contributions. First, it suggests a solution to the ongoing debate over which democratic institutions lead politicians to be the most responsive to special interests. Research on this topic has reached something of a stalemate. Theoretical models make competing predictions and empirical studies find conflicting results. Some studies find that particularistic policies are more frequent in plurality electoral systems. Yet others show that proportional electoral systems generate more particularistic economic policies than plurality systems. Although evidence continues to accumulate on both sides of the debate, no study has yet offered a solution for this impasse. I suggest a straightforward explanation for this gridlock: economic geography.

Most accounts of institutions' policy effect ignore economic geography. Others make highly restrictive assumptions about the geographic distribution of voters with shared policy preferences. For example, some scholars assume that voters employed in a given industry are entirely concentrated in a single electoral district (e.g. McGillivray 1997, Grossman and Helpman 2005). Yet, in reality, an industry's employees may be more or less geographically diffuse. As illustrated in Chapter 4, improved data transparency makes it possible to empirically measure the geographic distribution of voters employed in the same industry or sector. As a result, it is no longer necessary to make simplifying assumptions about where industries or their employees are located. Instead, their actual geographic distribution can be measured empirically – albeit for a limited number of countries.

Relaxing the restrictive assumptions about economic geography reveals new predictions about electoral institutions' policy effects. Both plurality and proportional systems generate particularistic economic policies in certain cases – it depends on the geographic concentration of voters with shared economic interests. By accounting for the variation exhibited in the spatial patterns of economic activity both within and between countries, my argument provides a bridge between two competing arguments and specifies the conditions under which one is more appropriate than the other.

In addition to taking economic geography seriously, my research also suggests several further ways to advance our understanding of the policy effects of electoral institutions. Understanding how electoral institutions – a fundamental feature of democracy – influence policy outcomes is important. Doing so brings new evidence to key questions at the heart of democratic politics including questions about responsiveness and representation. To advance our understanding of the policy effects of electoral institutions, I move beyond the blunt dichotomy between plurality and proportionality in Chapter 6 to explore the institutional variation that exists among PR countries. The findings reported in Chapter 6 suggest that the widespread use of the plurality/PR dichotomy obscures an important mechanism linking electoral institutions to policy: the nature of electoral competition. Electoral competition can be characterized as being either candidate-centered or party-centered. Party-centered competition encourages voters to emphasize their party preference over that for specific candidates. In contrast, candidate-centered competition encourages voters to see the basic unit of representation as the candidate rather than the party.

The PR/plurality dichotomy obscures the effects of electoral competition because the nature of electoral competition does not line up with the PR/plurality distinction. A country's electoral formula alone does not determine the nature of electoral competition. Instead, the constellation of institutions that make up a country's electoral system shape the nature of electoral competition. Proportional electoral formulas, for example, do not always generate party-centered elections. Elections in some PR systems are, in fact, candidate-centered. Open party lists generate candidate-centered electoral competition and elections in open-list PR systems are particularly candidate-centered when district magnitude is high. In sum, various features of a country's electoral system work together to shape the nature of electoral competition. This observation cautions against using only the blunt PR/plurality dichotomy to study the policy effects of electoral institutions.

Further confusion about the policy effects of electoral institutions stem from poor measures of particularistic economic policy. It is difficult to determine precisely which government programs should be classified as "particularistic" or narrowly-beneficial. Indeed, many existing studies struggle with this challenge (e.g. Cox and McCubbins 2001, Hatfield and Hauk 2014, Milesi-Ferretti, Perotti, and Rostagno 2002). My research suggests a novel way to classify government programs by looking at the number and geographic distribution of potential beneficiaries. In Chapter 4, I demonstrate how to estimate the geographic dispersion of an industry's employees using entropy indices. While this strategy is data intensive, it allows researchers to more accurately characterize government programs as being either "broad" or "narrow." Improving measures of particularistic economic policies using the strategy I employ here may help to clarify the policy effects of electoral institutions.

My argument makes two additional contributions of note. First, it adds an important new element to neoinstitutional theories in political science. These arguments contend that institutions aggregate preferences and acknowledge the importance of voters' preferences. However, it matters not only what voters want from government but also where they are located. Voters with shared preferences can be more or less geographically concentrated and their geographic patterns may change over time, particularly in an era of increased geographic mobility. Different electoral institutions create varied incentives for leaders to respond to geographically concentrated (or diffuse) groups. Therefore, geography must be part of any institutional story.

Second, my research contributes to understanding the political consequences of geographic concentration. Previous studies show that geographic concentration is a political asset in plurality systems, helping industries to win greater trade protection, for example. However, the role of geographic concentration in proportional systems remained unknown to date. This book provides one of the first quantitative test of the effects of geographic concentration between countries with varying electoral systems. The results reveal that the effects of geographic concentration vary across different electoral systems. In plurality systems, geographic concentration is a political asset. More concentrated groups win relatively more generous government subsidies, all else equal. However, geographic concentration is not always a political asset. In proportional systems, geographic concentration is, in fact, a political liability. Geographically concentrated groups win fewer subsidies than diffuse groups in PR systems, all else equal. This evidence challenges the long-held conventional wisdom that geographic concentration is politically valuable for interest groups.

BEYOND SUBSIDIES

I develop my argument in the context of economic policy and subsidies in particular. However, the logic of my argument is general and can be applied to a range of other issues. My argument provides useful insights whenever voters with shared preferences exhibit varied geographic patterns between and/or within countries. One example may be ethnic politics (Horowitz 1985, Mozaffar et al. 2003, Lijphart 2004). Members of an ethnic group may share common policy preferences. If so, my argument suggests that the influence of an ethnic group's preferences on policy will depend on the country's electoral institutions and the group's geographic distribution. When an ethnic group is geographically diffuse, their preferences will have greater expression under proportional electoral rules than plurality rules. In such a situation, it may be inopportune to introduce plurality electoral rules – particularly in an ethnically diverse society. Of course, future research is needed to determine the extent to which my argument applies to ethnic politics but it suggests one promising application beyond economic policy.

My argument cautions that institutions alone cannot guarantee a particular policy outcome. Policy is shaped by the interaction of institutions and geography and as a result, no specific institution can guarantee a desired policy outcome. Understanding that political institutions alone do not determine policy outcomes has important implications for constitutional designers and reformers. To correctly anticipate the policy effects of a particular institution, constitutional designers need to know the geographic patterns of politically relevant groups.

My argument also suggests a novel explanation for why government spending on social welfare tends to be higher in PR countries than plurality countries. In short, social welfare beneficiaries tend to be geographically diffuse (Persson and Tabellini 2003). Their geographic diffusion gives them greater political influence in PR countries as compared to plurality countries.[3] In this way, economic geography may help to explain why social welfare spending tends to be higher in PR systems than plurality systems – a question that has long intrigued many scholars.

Unemployed persons also tend to be geographically diffuse across a country. This pattern may help to explain why governments in PR countries tend to spend more on programs available to all unemployed persons, such as unemployment insurance, training, and job search assistance (Persson and Tabellini 2003). Leaders in plurality countries are not immune to concerns about unemployment. However, they tend to fund more selectively targeted unemployment programs. In the United States, for

---

[3] Others have suggested that this pattern emerges because PR systems favor left-leaning parties (Iversen and Soskice 2006, Rodden unpublished manuscript). While this mechanism may be at work, my argument suggests another possible mechanism: geography.

example, the federal government funds a program called Trade Adjustment Assistance (TAA) that aids people made unemployed by foreign trade (Rickard 2015). TAA benefits are selective and the eligibility criteria are strict. Workers must demonstrate that they lost their job as a direct result of foreign trade.[4] Workers displaced because of international trade tend to be geographically concentrated (OECD 2007). In fact, workers made unemployed by foreign trade are frequently more concentrated geographically than the population of unemployed persons (Autor et al. 2013). The geographic concentration of trade-displaced workers helps to explain why they receive selective unemployment benefits in countries with plurality electoral systems, like the United States. Governments in plurality systems have electoral incentives to assist geographically concentrated groups. As a result, they will augment general assistance programs with targeted, specific support programs when there are geographically concentrated job losses (OECD 2007). The benefits of the United States' TAA program, for example, go to geographically concentrated groups – often workers formerly employed in a single firm.[5] Although the United States' TAA program is relatively unique, other countries with candidate-centered elections also fund targeted assistance programs, including, for example, Mexico's Sectoral Promotion Programs (PROSEC) and Japan's System for Revitalizing Industrial Competitiveness. These programs selectively target aid to industries and regions where trade-induced unemployment is high. By funding assistance specifically for trade-displaced workers, leaders help geographically concentrated groups. For politicians in plurality systems, the electoral benefits from providing such targeted assistance are relatively large.

INTERNATIONAL IMPLICATIONS

Beyond implications for domestic politics, my argument also has important implications for international politics. First, my argument

---

[4] Consequently, only a small number of people benefit from the TAA program. In 2007, for example, the United States spent $855.1 million dollars to assist approximately 150,000 workers under the TAA program (Reynolds and Palatucci 2012) and in 2010, the program helped just 199,238 people (Dolfin and Schochet 2012).

[5] See Reynolds and Palatucci (2012) and https://doleta.gov/taaccct/pdf/TAACCCT_Maps_DH.pdf. Firms must first be certified as being eligible for the TAA program (Reynolds and Palatucci 2012). To be certified, groups of workers or their representatives file a petition with the Employment and Training Administration of the U.S. Department of Labor (USDOL). Once a firm is certified, workers laid off from the firm are then eligible for TAA benefits (Reynolds and Palatucci 2012).

suggests which countries are most likely to violate international economic agreements. Second, my argument identifies the countries most likely to impede future international economic integration. Third, my research identifies the countries most likely to demand reform to existing international economic agreements. I briefly discuss each implication in turn.

Violating International Economic Agreements

Some international economic agreements limit governments' ability to subsidize domestic producers. The WTO Agreement on Subsidies and Countervailing Measures (also known as the Subsidies Agreement or the SCM Agreement) establishes rules regulating the use of subsidies. The WTO's subsidy rules are enforceable through binding dispute settlement, which specifies strict time lines for bringing an offending program into conformity with WTO member-states' obligations (Davis 2012). The European Union's State Aid rules also limit certain types of subsidies. These rules are enforced by the European Commission, which can order a member state to recover aid granted to beneficiaries under an illegal subsidy scheme.[6]

In many countries, elected leaders are consequently caught between a rock and a hard place. They have domestic electoral incentives to provide subsidies but at the same time, international rules limit their ability to subsidize domestic producers. And violating international subsidy rules entails costs. Caught in this double-bind, what do leaders do? I argue that leaders will violate their international treaty obligations when the electoral benefits of doing so outweigh the costs. The electoral benefits are most likely to outweigh the costs when the incentives generated by a country's electoral institutions align with economic geography. As I have argued, subsidies bring large electoral benefits to leaders in plurality countries when the beneficiaries are geography concentrated. When beneficiaries are geographically diffuse, subsidies bring relatively greater benefits to parties in PR systems. When the incentives generated by a country's electoral institutions align with economic geography, leaders have powerful incentives to supply subsidies and these incentives may be sufficiently large to cause them to violate international subsidy rules.

---

[6] Recovery includes interest on the subsidy amount at an appropriate rate fixed by the Commission. Although these costs are substantial, the penalties for acting against EU subsidies rules are not always severe enough to deter "illegal" subsidies (Martin and Valbonesi 2008). On June 30, 2014, there were forty-nine active pending recovery cases.

This assertion is supported by evidence presented in Chapter 5, which shows that violations of EU state aid rules depend, in part, on a country's electoral institutions and economic geography. Politicians elected via proportional electoral rules are more likely to subsidize geographically diffuse sectors in violation of EU state aid rules than politicians elected via plurality rules. In other words, governments in PR systems provide more "illegal" (i.e. non-EU compliant) subsidies than governments in plurality systems when the beneficiaries are geographically diffuse. Further evidence of this pattern comes from the Austrian subsidy for farm-gate wine merchants discussed in Chapter 5. The Austrian legislature, whose members are elected via proportional rules from de facto closed party lists, supplied assistance to farm-gate wine merchants spread across all of the country's winemaking regions. The Austrian government violated EU state aid rules to subsidize this geographically diffuse group because the electoral benefits of doing so were large.

Resistance to New International Rules

Governments often have electoral incentives to subsidize domestic producers. Yet, compelling reasons exist to agree to international restrictions on subsidies. Countries are better off if they can reach and enforce an agreement to forgo subsidies funded in the prospect of poaching each other's profits in imperfectly competitive markets (Dewatripont and Seabright 2006).[7] Agreeing to international subsidy restrictions may help countries avoid wasteful, and often escalating, "subsidy wars" (Dewatripont and Seabright 2006). Subsidy wars can be costly and protracted, as illustrated by the experience of the large civil aircraft industry. This industry is dominated by two firms: Airbus and Boeing. Airbus was created by a group of European nations to rival the American firm Boeing. Airbus received generous financial assistance from its' member nations. They justified their financial assistance by citing the high barriers to entry in the capital-intensive aircraft industry (Kienstra 2012). Due in part to the European governments' largesse, Airbus eventually overtook Boeing's position as the leading producer of large civil aircraft (Kienstra 2012). The United States government responded by providing generous financial assistance to Boeing.

---

[7] In fact, the logic behind international restrictions on subsidies builds on "strategic trade policy" where countries compete with each other in a game of individually rational but collectively wasteful subsidies to industry, spurred by the prospect of poaching each other's profits in imperfectly competitive markets (Brander and Spencer 1985).

The firm received an array of subsidies from different government agencies including America's space agency, NASA, and the Export-Import Bank of the United States (sometimes referred to by critics as "Boeing's Bank"). These subsidies brought the United States into conflict with the European Community, which reached a head in 2004 when the United States filed a formal complaint with the World Trade Organization over subsidies provided by the European Community to Airbus. The European Community responded by filing a parallel complaint regarding subsidies provided to Boeing by the United States. The European Community claimed that Boeing benefited from illegal subsidies worth $19.1 billion between 1989 and 2006.[8] The dispute remains ongoing; the most recent action, a panel report, was circulated in September 2016. As this case illustrates, subsidy wars can be costly and drawn out. Countries therefore have incentives to agree to (and comply with) international rules regulating subsidies.

These international incentives, however, must be weighed against the domestic electoral benefits from providing subsidies. Governments may want to provide subsidies to domestic producers in order to reap the electoral benefits. In this case, governments may resist further international restrictions on subsidies. More international rules mean governments will have fewer opportunities to use subsidies to win elections. As shown in Chapter 5, governments may flout international rules and provide subsidies in violation of international economic agreements when the electoral benefits of doing so are large. But violating international rules is costly. The Italian government was ordered to recover illegal aid amounting to €15 million from hotel companies in Sardinia by the European Commission in 2008. Seven years after the order, only €2 million of the initial €15 had been recovered. In 2015, the Commission asked the European Court of Justice to impose a lump sum penalty of €20 million, in addition to a daily penalty payment of €160,000 until Italy fully recovered the aid (European Commission 2015). As this example shows, violating international restrictions on subsidies can be very costly. Governments would generally prefer to avoid these costs, and as a result they may actively resist new international restrictions on subsidies.

Governments with the greatest electoral incentives to supply subsidies will put up the most resistance to new international restrictions on subsidies. Governments elected via plurality rules, for example, will oppose new international restrictions on subsidies to geographically concentrated groups, as illustrated by Canada's intransigence in

---

[8] WTO Dispute DS353, www.wto.org/english/tratop_e/dispu_e/cases_e/ds353_e.htm.

negotiations over the Trans-Pacific Partnership (TPP) agreement. In these negotiations, twelve countries worked for five years to reach an agreement that would cover nearly 40 percent of global trade. At the very last minute, subsidies to fewer than 13,000 Canadian dairy farms nearly sank the deal.[9] The disagreement arose over dairy subsidies between Canada and dairy-exporting countries, such as New Zealand, who wanted greater access to Canada's market. Canada strongly defended their supply management program, which heavily subsidizes domestic dairy farmers. The small group of Canadian dairy producers enjoyed disproportionate political influence because of their geographic concentration and Canada's electoral institutions. Dairy farmers are chiefly concentrated in the Canadian provinces of Ontario and Quebec. Almost half (49 percent) of Canada's milk farms are in Quebec, and neighboring Ontario has 32 percent of Canada's dairy producing farms (Andrew-Gee 2015, Canadian Dairy Information Centre 2015).[10] By defending dairy subsides, the Prime Minister sought to bolster his party's electoral support in Ontario and Quebec. In Canada, politicians are elected to office via plurality electoral rules in single-member districts. In such a system, where elections are won district-by-district and province-by-province, politicians have incentives to provide subsidies to producers concentrated in key areas. The Prime Minister sought to increase his party's votes by targeting aid to dairy farmers in Ontario and Quebec – even at the risk of being left out of the TPP agreement. Although Canada ultimately won concessions for the domestic dairy industry, the Canadian government also promised to compensate dairy farmers for any losses they incurred because of TPP.

Similarly, France objected to subsidy restrictions proposed as part of a new trade agreement between the EU and the United States known as the Transatlantic Trade and Investment Partnership (TTIP). French leaders threatened to kill TTIP talks before they even began, in opposition to anticipated limits on state subsidies.[11] At the heart of France's opposition were subsidies to the audiovisual services industry,

---

[9] According to New Zealand's trade minister, negotiations on dairy subsidies only ended around 5 a.m. on October 5, 2015. (Press conference, Monday, October 5, 2015, Atlanta, Georgia). See also, "Things to know about Canada's dairy supply management system." CTV News. *The Canadian Press*, published Sunday, October 4, 2015, www.ctvnews.ca/business/things-to-know-about-canada-s-dairy-supply-management-system-1.2594529.

[10] www.dairyinfo.gc.ca/index_e.php?s1=dff-fcil&s2=farm-ferme&s3=nb.

[11] The Economist, Charlemagne, "L'exception française, A transatlantic free-trade deal is needlessly held up over subsidies for film-makers," June 15, 2013.

which is concentrated in the Île-de-France region (Dale 2015). The industry's geographic concentration contributes to its political clout (Frey 2014). The industry won protectionist amendments to the Franco-American trade agreement (Frey 2014) and today benefits from subsidies worth nearly €1 billion a year (Carnegy 2013). To maintain this support, France demanded that subsidies to the industry be excluded from any new restrictions on state aid agreed as part of the Transatlantic Trade and Investment Partnership (TTIP). In sum, electoral institutions and economic geography help to explain which countries resist further international restrictions on government-funded subsidies.

Demands for Reform

Governments that frequently run afoul of international restrictions on subsidies have the most to gain from reform. Given this, I speculate that the same variables that incentivize politicians to provide "illegal" subsidies will also lead them to demand reform to existing international rules. Leaders in plurality-rule countries who face geographically concentrated producers, for example, have powerful domestic incentives to demand reform to international subsidy rules. Indeed, French leaders lobbied for changes to EU rules that restrict government-funded subsidies (i.e. state aid). In 2014, the French industry minister launched a blistering attack on EU state aid rules calling them "obsolete" (Carnegy and Stothard 2014). France's industry minister argued that the EU's strict application of controls on state subsidies was preventing Europe from competing in the global market. The minister complained, "While our global industrial competitors get billions in subsidies, our bureaucracy is led by political leaders...... It is like Rome surrounded by the barbarians. We all await the fall of Rome. It's not funny" (Carnegy and Stothard 2014). The UK joined France in demanding reform to the European Union's state aid rules. These demands came in response to increased competition from China, the United States, and Japan. Some European leaders believe the EU subsidy rules hinder member-state governments' attempts to assist European companies. "This is how the European Commission is working to weaken our industry. Do you sincerely believe that this is reasonable?" Mr. Montebourg, the French industry minister wrote regarding EU state aid rules (Carnegy and Stothard 2014).

While France and the United Kingdom lobbied for changes to state aid rules, other EU member-states defend the existing restrictions. My argument helps to explain which countries push for reform to international subsidy rules and why. Governments in plurality systems like France and the United Kingdom will lobby for a relaxation of EU state aid rules to subsidize politically important, geographically concentrated groups. France, for example, focused special attention on exempting subsidies to the geographically concentrated film industry. The UK government joined France in supporting exemptions for the film industry. Like the French industry, the British film industry is geographically concentrated. The production, postproduction, and distribution sectors are clustered in London and the South East of England (BFI Statistical Yearbook 2015). The efforts of France and the UK on behalf of their respective film industries were ultimately successful. Just months after the French government threatened to veto the bloc's trade talks with the United States, the European Commission published new rules making it easier for governments to subsidize moviemaking (Fox 2013). Under the new rules, governments will be allowed to cover 50 percent of the costs of a film from scriptwriting and production to distribution and promotional costs (Fox 2013). Governments will also be able to require that between 50 to 80 percent of subsidized films' budgets be spent within the country (Fox 2013). In a joint statement, UK Chancellor of the Exchequer George Osborne together with the British Film Institute said that the reforms to the EU subsidy rules were a "huge reassurance to the UK film industry," adding that the UK's film sector was responsible for 117,000 jobs (Fox 2013).

General demands for reform to EU state aid rules were also successful. In May 2014, the European Commission made it easier for member-states to assist companies using subsidies. Firms can now benefit from subsidies for a broader range of activities. Under the new rules, subsidies are allowable when the aid is clearly aimed at creating jobs or boosting competitiveness. Governments can also provide higher amounts of financial assistance without their plans being subject to prior scrutiny by the EU authorities. The reforms to EU state aid rules give member-states greater freedom to subsidize business.

As EU restrictions on state aid are relaxed, the share of governments' budgets devoted to subsidies will likely increase. Government spending on subsidies may further increase in response to pressures to assist the declining manufacturing sector. Increased government spending on subsidies has important consequences. Money spent on subsidies is money that is no longer available for other programs, such as education, health care, and social welfare. Governments facing tight budget constraints often cut spending on programs, such as social welfare, to fund industrial subsidies (Rickard 2012b). Subsidies consequently have serious implications for the regressivity of government spending. It is therefore important to understand why leaders in some democracies are willing to devote larger shares of their budgets to subsidies. Political institutions alone cannot explain this choice. Instead, economic geography must be considered together with electoral institutions to fully understand governments' economic policy decisions.

References

Aardal, Bernt. 1990. "The Norwegian Parliamentary Election of 1989." *Electoral Studies* 9 (2): 151–158.
   2011. "The Norwegian Electoral System and Its Political Consequences." *World Political Science* 7 (1). doi:10.2202/1935-6226.1105.
Aberbach, Joel D. 1990. *Keeping a Watchful Eye: The Politics of Congressional Oversight*. Brookings Institution Press: Washington D.C.
Aghion, Philippe, Mathias Dewatripont, Luosha Du, Ann E. Harrison, and Patrick Legros. 2011. "Industrial Policy and Competition." SSRN Scholarly Paper ID 1811643. Social Science Research Network. http://papers.ssrn.com/abstract=1811643.
Aidt, Toke S., Miriam A. Golden, and Devesh Tiwari. 2011. "Incumbents and Criminals in the Indian National Legislature." Working Paper. Faculty of Economics. www.repository.cam.ac.uk/handle/1810/242058.
Aiken, Leona S., Stephen G. West, and Raymond R. Reno. 1991. *Multiple Regression: Testing and Interpreting Interactions*. Newbury Park, Calif.: Sage Publications.
Alesina, Alberto, Nouriel Roubini, and Gerald D. Cohen.1997. *Political Cycles and the Macroeconomy*. Cambridge, Mass.: MIT Press.
Allen, Will. 2004. "Fact Sheet on U.S. Cotton Subsidies and Cotton Production." www.organicconsumers.org/old_articles/clothes/224subsidies.php.
Alt, James E., Fredrik Carlsen, Per Heum, and Kåre Johansen. 1999. "Asset Specificity and the Political Behavior of Firms: Lobbying for Subsidies in Norway." *International Organization* 53 (1): 99–116. doi:10.1162/002081899550823.
Alt, James E. and Michael Gilligan. 1994. "The Political Economy of Trading States: Factor Specificity, Collective Action Problems and Domestic Political Institutions." *Journal of Political Philosophy* 2 (2): 165–192. doi:10.1111/j.1467-9760.1994.tb00020.x.
Alvarez, R. Michael and Jason L. Saving. 1997. "Deficits, Democrats, and Distributive Benefits: Congressional Elections and the Pork Barrel in the 1980s." *Political Research Quarterly* 50 (4): 809–831. doi:10.1177/106591299705000405.
Ames, Barry. 1995. "Electoral Strategy under Open-List Proportional Representation." *American Journal of Political Science* 39 (2): 406–433. doi:10.2307/2111619.

Anderson, Christopher J. and Pablo Beramendi. 2012. "Left Parties, Poor Voters, and Electoral Participation in Advanced Industrial Societies." *Comparative Political Studies* 45 (6): 714–746. doi:10.1177/0010414011427880.

Anderson, Kym. 2009. "Distorted Agricultural Incentives and Economic Development: Asia's Experience." *The World Economy* 32(3): 351–84.

Anderson, Kym and Johan F. M. Swinnen. 2008. *Distortions to Agricultural Incentives in Europe's Transition Economies*. Washington, DC: World Bank. http://public.eblib.com/choice/publicfullrecord.aspx?p=459326.

Andersson, David Emanuel, Åke E. Andersson, Björn Hårsman, and Zara Daghbashyan. 2015. "Unemployment in European Regions: Structural Problems versus the Eurozone Hypothesis." *Journal of Economic Geography* 15 (5): 883–905. doi:10.1093/jeg/lbu058.

Andrew-Gee, Eric. 2015. "Anxiety Pervades Quebec's Dairy Farms as TPP Talks Heat Up." *The Globe and Mail*, July 31. www.theglobeandmail.com/news/national/anxiety-pervades-quebecs-dairy-farms-as-tpp-talks-heat-up/article25810043/.

Ansolabehere, Stephen, Alan Gerber, and James Snyder. 2002. "Equal Votes, Equal Money: Court-Ordered Redistricting and Public Expenditures in the American States." *The American Political Science Review* 96 (4): 767–777.

Arbia, Giuseppe. 2001. "Modelling the geography of economic activities on a continuous space." *Papers in Regional Science* 80(4): 411-424.

Ardelean, Adina and Carolyn L. Evans. 2013. "Electoral Systems and Protectionism: An Industry-Level Analysis." *Canadian Journal of Economics/Revue Canadienne D'économique* 46 (2): 725–764. doi:10.1111/caje.12030.

Arnold, R. Douglas. 1990. *The Logic of Congressional Action*. New Haven, Conn.: Yale University Press.

Ásgeirsdóttir, Áslaug. 2008. *Who Gets What: Domestic Influences on International Negotiations Allocating Shared Resources*. Albany, N.Y.: SUNY Press. http://public.eblib.com/choice/publicfullrecord.aspx?p=3407415.

Ashworth, Scott and Ethan Bueno de Mesquita. 2006. "Delivering the Goods: Legislative Particularism in Different Electoral and Institutional Settings." *Journal of Politics* 68 (1): 168–79. doi:10.1111/j.1468-2508.2006.00378.x.

Austrian Wine Marketing Board. 2016. "Austrian Wine Statistics Report." www.austrianwine.com. Accessed July 13, 2017. www.austrianwine.com/facts-figures/austrian-wine-statistics-report/.

Autor, David, David Dorn, Gordon Hanson, and Kaveh Majlesi. 2016. "Importing Political Polarization? The Electoral Consequences of Rising Trade Exposure." NBER Working Paper No. 22637. Issued in September 2016.

Autor, David H., David Dorn, and Gordon H. Hanson. 2013. "The Geography of Trade and Technology Shocks in the United States." *The American Economic Review* 103 (3): 220–225. doi:10.1257/aer.103.3.220.

Aydin, Umut. 2007. "Promoting Industries in the Global Economy: Subsidies in OECD Countries, 1989 to 1995." *Journal of European Public Policy* 14 (1): 115–131. doi:10.1080/13501760601071976.

Bailey, Alan. 2013. "Not All 'the Bad Old Days': Revisiting Labour's 1970s Industrial Strategy." www.ippr.org/juncture/not-all-the-bad-old-days-revisiting-labours-1970s-industrial-strategy.

Bailey, Michael. 2001. "Quiet Influence: The Representation of Diffuse Interests on Trade Policy, 1983–94." *Legislative Studies Quarterly* 26 (1): 45–80. doi:10.2307/440403.

Balla, Steven J., Eric D. Lawrence, Forrest Maltzman, and Lee Sigelman. 2002. "Partisanship, Blame Avoidance, and the Distribution of Legislative Pork." *American Journal of Political Science* 46 (3): 515–25. doi:10.2307/3088396.

Ballard-Rosa, Cameron, Stephanie J. Rickard, and Ken Scheve. 2017. "Liberal Populism: Public Support for Globalization in Post-Brexit United Kingdom." Working paper. University of North Carolina.

Banks, Jeffrey S. 1989. "Agency Budgets, Cost Information, and Auditing." *American Journal of Political Science* 33 (3): 670–699. doi:10.2307/2111068.

Barber, Benjamin. 2014. "The Political Economy of Decline." Duke University. PhD thesis.

Barkan, Joel D., Paul J. Densham, and Gerard Rushton. 2006. "Space Matters: Designing Better Electoral Systems for Emerging Democracies." *American Journal of Political Science* 50 (4): 926–939. doi:10.1111/j.1540-5907.2006.00224.x.

Bawn, Kathleen and Frances Rosenbluth. 2006. "Short versus Long Coalitions: Electoral Accountability and the Size of the Public Sector." *American Journal of Political Science* 50 (2): 251–65. doi:10.1111/j.1540-5907.2006.00182.x.

Beattie, Alan. 2014. "The US and India: A Trade Truce with a Twist." *Financial Times*. November 19. http://blogs.ft.com/beyond-brics/2014/11/19/the-us-and-india-a-trade-truce-with-a-twist/. Accessed March 13, 2017.

BBC. 2009. "Saab Board Ends Emergency Meeting," February 19, sec. Business. http://news.bbc.co.uk/1/hi/business/7899244.stm.

Beck, Thorsten, George Clarke, Alberto Groff, Philip Keefer, and Patrick Walsh. 2001. "New Tools in Comparative Political Economy: The Database of Political Institutions." *The World Bank Economic Review* 15 (1): 165–176. doi:10.1093/wber/15.1.165.

Becker, Gary S. 1985. "Public Policies, Pressure Groups, and Dead Weight Costs." *Journal of Public Economics* 28 (3): 329–347. doi:10.1016/0047-2727(85)90063-5.

Beghin, John C., Barbara El Osta, Jay R. Cherlow, and Samarendu Mohanty. 2003. "The Cost of the U.S. Sugar Program Revisited." *Contemporary Economic Policy* 21 (1): 106–16. doi:10.1093/cep/21.1.106.

Bendor, Jonathan, Serge Taylor, and Roland Van Gaalen. 1987. "Politicians, Bureaucrats, and Asymmetric Information." *American Journal of Political Science* 31 (4): 796–828. doi:10.2307/2111225.

Beramendi, Pablo. 2012. *The Political Geography of Inequality: Regions and Redistribution*. New York, N.Y.: Cambridge University Press.

Bernanke, Ben, Kenneth Rogoff, and National Bureau of Economic Research, eds. 2001. *NBER Macroeconomics Annual 2000*. Cambridge, Mass.: MIT Press.

Besley, Timothy and Paul Seabright. 1999. "The Effects and Policy Implications of State Aids to Industry: An Economic Analysis." *Economic Policy* 14 (28): 14–53. doi:10.1111/1468-0327.00043.

BFI. 2015. "Statistical Yearbook." *British Film Institute*. www.bfi.org.uk/education-research/film-industry-statistics-research/statistical-yearbook.

Blais, André. 1986. "The Political Economy of Public Subsidies." *Comparative Political Studies* 19 (2): 201–216. doi:10.1177/0010414086019002002.
Blais, André and Louis Massicotte. 1997. "Electoral Formulas: A Macroscopic Perspective." *European Journal of Political Research* 32 (1): 107–129. doi:10.1023/A:1006839007213.
Blauberger, Michael. 2009. "Of 'Good' and 'Bad' Subsidies: European State Aid Control through Soft and Hard Law." *West European Politics* 32 (4): 719–737. doi:10.1080/01402380902945300.
Blomström, Magnus, Ari Kokko, and Jean-Louis Mucchielli. 2003. "The Economics of Foreign Direct Investment Incentives," in *Foreign Direct Investment in the Real and Financial Sector of Industrial Countries*, edited by Heinz Herrmann and Robert Lipsey, 37–60. Springer Berlin Heidelberg. http://link.springer.com/chapter/10.1007/978-3-540-24736-4_3.
Boix, Carles. 1999. "Setting the Rules of the Game: The Choice of Electoral Systems in Advanced Democracies." *American Political Science Review* 93 (3): 609–624. doi:10.2307/2585577.
Bown, Chad P. 2017. "Trump's Threat of Steel Tariffs Heralds Big Changes in Trade Policy," *Washington Post*. April 21, 2017.
Brambor, Thomas, William Roberts Clark, and Matt Golder. 2006. "Understanding Interaction Models: Improving Empirical Analyses." *Political Analysis* 14 (1): 63–82. doi:10.1093/pan/mpi014.
Brander, James A. and Barbara J. Spencer. 1985. "Export Subsidies and International Market Share Rivalry." *Journal of International Economics* 18 (1): 83–100. doi:10.1016/0022-1996(85)90006-6.
Braunerhjelm, Pontus and Benny Borgman. 2004. "Geographical Concentration, Entrepreneurship and Regional Growth: Evidence from Regional Data in Sweden, 1975–99." *Regional Studies* 38 (8): 929–947. doi:10.1080/0034340042000280947.
Brouard, Sylvain, Olivier Costa, Eric Kerrouche, and Tinette Schnatterer. 2013. "Why Do French MPs Focus More on Constituency Work than on Parliamentary Work?" *The Journal of Legislative Studies* 19 (2): 141–159. doi:10.1080/13572334.2013.787194.
Broz, J. Lawrence. 2005. "Congressional Politics of International Financial Rescues." *American Journal of Political Science* 49 (3): 479–96. doi:10.1111/j.1540-5907.2005.00137.x.
Brülhart, Marius and Rolf Traeger. 2005. "An Account of Geographic Concentration Patterns in Europe." *Regional Science and Urban Economics* 35 (6): 597–624. doi:10.1016/j.regsciurbeco.2004.09.002.
Brunori, David. 2014. "Where Is the Outrage Over Corporate Welfare?" *Forbes*. Accessed July 26, 2107 www.forbes.com/sites/taxanalysts/2014/03/14/where-is-the-outrage-over-corporate-welfare/.
Buenstorf, Guido and Christina Guenther. 2010. "No Place like Home? Relocation, Capabilities, and Firm Survival in the German Machine Tool Industry after World War II." *Industrial and Corporate Change*, September, dtq055. doi:10.1093/icc/dtq055.
Busch, Marc L. and Eric Reinhardt. 2000. "Geography, International Trade, and Political Mobilization in U.S. Industries." *American Journal of Political Science* 44 (4): 703–719. doi:10.2307/2669276.

2005. "Industrial Location and Voter Participation in Europe." *British Journal of Political Science* 35 (4): 713–730. doi:10.1017/S0007123405000360.
Buts, Caroline, Marc Jegers, and Dimi Jottier. 2012. "The Effect of Subsidising Firms on Voting Behaviour: Evidence from Flemish Elections." *European Journal of Government and Economics* 1 (1): 30–43.
Calvert, Randall L., Mathew D. McCubbins, and Barry R. Weingast. 1989. "A Theory of Political Control and Agency Discretion." *American Journal of Political Science* 33 (3): 588–611. doi:10.2307/2111064.
Calvo, Ernesto and Maria Victoria Murillo. 2004. "Who Delivers? Partisan Clients in the Argentine Electoral Market." *American Journal of Political Science* 48 (4): 742–57. doi:10.1111/j.0092-5853.2004.00099.x.
Campos, Cecilia. 2012. "The Geographical Concentration of Industries." United Kingdom: Office of National Statistics. http://webarchive.nationalarchives.gov.uk/20160105160709/www.ons.gov.uk/ons/dcp171766_272232.pdf.
Carey, John M. and Simon Hix. 2011. "The Electoral Sweet Spot: Low-Magnitude Proportional Electoral Systems." *American Journal of Political Science* 55 (2): 383–97. doi:10.1111/j.1540-5907.2010.00495.x.
Carey, John M. and Matthew Soberg Shugart. 1995. "Incentives to Cultivate a Personal Vote: A Rank Ordering of Electoral Formulas." *Electoral Studies* 14 (4): 417–439. doi:10.1016/0261-3794(94)00035-2.
Carlsson, Bo. 1983. "Industrial Subsidies in Sweden: Macro-Economic Effects and an International Comparison." *The Journal of Industrial Economics* 32 (1): 1–23. doi:10.2307/2097983.
Carnegy, Hugh. 2013. "France Defends Cultural Barricades in Digital Era." *Financial Times*, May 13, 2013. https://www.ft.com/content/b60f4182-bbea-11e2-a4b4-00144feab7de
Carnegy, Hugh and Michael Stothard. 2014. "French Industry Minister Assails Brussels on State Aid for Industry." *Financial Times*, January 22, 2014. www.ft.com/cms/s/0/99515d88-8367-11e3-86c9-00144feab7de.html#axzz4H2MCuSJq.
Cassing, James, Timothy J. McKeown, and Jack Ochs. 1986. "The Political Economy of the Tariff Cycle." *American Political Science Review* 80 (3): 843–862. doi:10.2307/1960541.
Chakrabortty, Aditya. 2015. "The £93bn Handshake: Businesses Pocket Huge Subsidies and Tax Breaks." *The Guardian*, July 7, sec. Politics. www.theguardian.com/politics/2015/jul/07/corporate-welfare-a-93bn-handshake.
Chang, Eric C. C. and Miriam A. Golden. 2007. "Electoral Systems, District Magnitude and Corruption." *British Journal of Political Science* 37 (1): 115–137. doi:10.1017/S0007123407000063.
Chang, Eric C. C., Mark Andreas Kayser, Drew Linzer, and Ronald Rogowski. 2010. *Electoral Systems and the Balance of Consumer-Producer Power*. Leiden: Cambridge University Press. http://public.eblib.com/choice/publicfullrecord.aspx?p=615739.
Chang, Eric C. C., Mark Andreas Kayser, and Ronald Rogowski. 2008. "Electoral Systems and Real Prices: Panel Evidence for the OECD Countries, 1970–2000." *British Journal of Political Science* 38 (4): 739–751.
Chase, Kerry. 2015. "Domestic Geography and Policy Pressures," in *The Oxford Handbook of the Political Economy of International Trade*, edited by Lisa Martin, 316–334. New York, N.Y.: Oxford University Press.

Chassany, Anne-Sylvaine. 2016. "France to buy unneeded trains to save Belfort factory." *Financial Times*. October 4, 2016. www.ft.com/content/9e7deeee-8a07-11e6-8aa5-f79f5696c731.

Cini, Michelle. 2001. "The Soft Law Approach: Commission Rule-Making in the EU's State Aid Regime." *Journal of European Public Policy* 8 (2): 192–207. doi:10.1080/13501760110041541.

Clark, William Roberts, Matt Golder, and Sona Nadenichek Golder. 2013. *Principles of Comparative Politics*. Thousand Oaks, CA: Sage.

Colantone, Italo and Piero Stanig. 2016. "Globalisation and Brexit." In Press.

Costa, Olivier and Eric Kerrouche. 2009. "Representative Roles in the French National Assembly: The Case for a Dual Typology?" *French Politics* 7 (3–4): 219–242. doi:10.1057/fp.2009.16.

Costa-i-Font, Joan, Eduardo Rodriguez-Oreggia, and Dario Lunapla. 2003. "Political Competition and Pork-Barrel." *Public Choice* 116 (1–2): 185–204. doi:10.1023/A:1024263208736.

Cowell, Frank A. 1980. "On the Structure of Additive Inequality Measures." *The Review of Economic Studies* 47 (3): 521–531. doi:10.2307/2297303.

1995. *Measuring Inequality*. London; New York, N.Y.: Prentice Hall/Harvester Wheatsheaf.

2000. "Chapter 2 Measurement of Inequality," in *Handbook of Income Distribution*, 1, edited by A.B. Atkinson and F. Bourguignon, 87–166. Elsevier. www.sciencedirect.com/science/article/pii/S1574005600800056.

Cox, Gary W. 1987. "Electoral Equilibrium under Alternative Voting Institutions." *American Journal of Political Science* 31 (1): 82–108. doi:10.2307/2111325.

1990. "Centripetal and Centrifugal Incentives in Electoral Systems." *American Journal of Political Science* 34 (4): 903–35. doi:10.2307/2111465.

1997. *Making Votes Count: Strategic Coordination in the World's Electoral Systems*. New York: Cambridge University Press. http://dx.doi.org/10.1017/CBO9781139174954.

Cox, Gary W. and Mathew D. McCubbins. 1986. "Electoral Politics as a Redistributive Game." *The Journal of Politics* 48 (2): 370–389. doi:10.2307/2131098.

1993. *Legislative Leviathan: Party Government in the House*. Berkeley, Calif.: University of California Press.

2001. "The Institutional Determinants of Economic Policy Outcomes," in *Presidents, Parliaments, and Policy*, edited by Stephan Haggard and Mathew D. McCubbins, 21–63. Cambridge: Cambridge University Press.

Cox, Gary W. and Frances Rosenbluth. 1993. "The Electoral Fortunes of Legislative Factions in Japan." *American Political Science Review* 87 (3): 577–589. doi:10.2307/2938737.

Cox, Karen E. and Leonard J. Schoppa. 2002. "Interaction Effects in Mixed-Member Electoral Systems Theory and Evidence From Germany, Japan, and Italy." *Comparative Political Studies* 35 (9): 1027–1053. doi:10.1177/001041402237505.

Creative Screen Associates. 2013. "Comparison of French and UK Public Policy and Support Mechanisms for the Film Industry." www.bectu.org.uk/advice-resources/library/1416.

Criscuolo, Chiara, Ralf Martin, Henry Overman, and John Van Reenen. 2012. "The Causal Effects of an Industrial Policy." Working Paper 17842. National Bureau of Economic Research. www.nber.org/papers/w17842.

Crisp, Brian F., Maria C. Escobar-Lemmon, Bradford S. Jones, Mark P. Jones, and Michelle M. Taylor-Robinson. 2004. "Vote-Seeking Incentives and Legislative Representation in Six Presidential Democracies." *Journal of Politics* 66 (3): 823–846. doi:10.1111/j.1468-2508.2004.00278.x.

Cusack, Thomas R., Torben Iversen, and David Soskice. 2007. "Economic Interests and the Origins of Electoral Systems." *American Political Science Review* 101 (3): 373–391. doi:10.1017/S0003055407070384.

Dale, Martin. 2015. "French Equipment Manufacturer Association – AFFECT – Attends the AFC Micro-Salon." *Variety*. February 5. http://variety.com/2015/film/news/french-equipment-manufacturer-association-affect-attends-the-afc-micro-salon-1201426072/.

Database, EWG's Farm Subsidy. 2016. "EWG's Farm Subsidy Database." Accessed July 26, 2017 http://farm.ewg.org/progdetail.php?fips=00000&progcode=cotton&yr=2014&page=district&regionname=theUnitedStates.

Davis, Christina L. 2012. *Why Adjudicate? Enforcing Trade Rules in the WTO*. Princeton, N.J.: Princeton University Press.

de Mesquita, Bruce Bueno, Alastair Smith, Randolph Siverson, and James D. Morrow. 2003. *The Logic of Political Survival*. Cambridge, Mass.: MIT Press.

Denemark, David. 2000. "Partisan Pork Barrel in Parliamentary Systems: Australian Constituency-Level Grants." *The Journal of Politics* 62 (3): 896–915. doi:10.1111/0022-3816.00039.

Denzau, Arthur T. and Michael C. Munger. 1986. "Legislators and Interest Groups: How Unorganized Interests Get Represented." *American Political Science Review* 80 (1): 89–106. doi:10.2307/1957085.

Depauw, Sam and Shane Martin. 2008. "Legislative Party Discipline and Cohesion in Comparative Perspective," in *Intra-Party Politics and Coalition Governments*, edited by Daniela Giannetti and Kenneth Benoit. Abington U.K.: Routledge.

Devereux, Michael P., Rachel Griffith, and Helen Simpson. 2007. "Firm Location Decisions, Regional Grants and Agglomeration Externalities." *Journal of Public Economics* 91 (3–4): 413–435. doi:10.1016/j.jpubeco.2006.12.002.

Dewatripont, Mathias, and Paul Seabright. 2006. "'Wasteful' Public Spending and State Aid Control." *Journal of the European Economic Association* 4 (2–3): 513–522. doi:10.1162/jeea.2006.4.2-3.513.

Dickson, Vaughan. 2009. "Seat-Vote Curves, Loyalty Effects and the Provincial Distribution of Canadian Government Spending." *Public Choice* 139 (3–4): 317–333. doi:10.1007/s11127-009-9395-1.

Dixit, Avinash and John Londregan. 1996. "The Determinants of Success of Special Interests in Redistributive Politics." *The Journal of Politics* 58 (4): 1132–1155. doi:10.2307/2960152.

Dolfin, Sarah and Peter Schochet. 2012. "The Benefits and Costs of the Trade Adjustment Assistance (TAA) Program under the 2002 Amendments." Princeton, N.J.: Mathematica Policy Research. www.mathematica-mpr.com

/our-publications-and-findings/publications/the-benefits-and-costs-of-the-tra de-adjustment-assistance-taa-program-under-the-2002-amendments.

Duchêne, François and Geoffrey Shepherd. 1987. *Managing Industrial Change in Western Europe*. London; New York, N.Y.: F. Pinter.

Dumais, Guy, Glenn Ellison, and Edward L. Glaeser. 2002. "Geographic Concentration as a Dynamic Process." *Review of Economics and Statistics* 84 (2): 193–204. doi:10.1162/003465302317411479.

Dunleavy, Patrick. 1991. *Democracy, Bureaucracy, and Public Choice: Economic Explanations in Political Science*. New York, N.Y.: Prentice Hall.

Duverger, Maurice. 1964. *An Introduction to the Social Sciences, with Special Reference to Their Methods*. New York, N.Y.: F.A. Praeger.

Ehrlich, Sean D. 2007. "Access to Protection: Domestic Institutions and Trade Policy in Democracies." *International Organization* 61 (3): 571–605. doi:10.1017/S0020818307070191.

Elder, Neil and Rolf Gooderham. 1978. "The Centre Parties of Norway and Sweden" *Government and Opposition* 13 (2): 218–35. doi:10.1111/j.1477-7053.1978.tb00544.x.

Elgie, Robert. 2005. "France: Stacking the Deck," in *The Politics of Electoral Systems*, edited by Michael Gallagher and Paul Mitchell, 119–138. Oxford University Press.

Ellison, Glenn, and Edward L. Glaeser. 1997. "Geographic Concentration in U.S. Manufacturing Industries: A Dartboard Approach." *Journal of Political Economy* 105 (5): 889–927. doi:10.1086/262098.

1999. "The Geographic Concentration of Industry: Does Natural Advantage Explain Agglomeration?" *The American Economic Review* 89 (2): 311–316.

Ellison, Glenn, Edward L. Glaeser, and William R. Kerr. 2010. "What Causes Industry Agglomeration? Evidence from Coagglomeration Patterns." *The American Economic Review* 100 (3): 1195–1213. doi:10.1257/aer.100.3.1195.

Epstein, David and Sharyn O'Halloran. 1994. "Administrative Procedures, Information, and Agency Discretion." *American Journal of Political Science* 38 (3): 697–722. doi:10.2307/2111603.

Euromonitor International. 2015. "Wine in Austria, Category Briefing." Oct. 12, 2015.

European Commission. 1995. "XXVth REPORT on Competition Policy." http://ec.europa.eu/competition/publications/annual_report/1995/en.pdf.

1997. "XXVIIth Report on Competition Policy." http://ec.europa.eu/competition/publications/annual_report/1997/broch97_en.pdf.

2003. *Raising EU R&D Intensity*. EUR Community Research 20717. Luxembourg: Off. for Official Publ. of the Europ. Communities.

2009. "Scoreboard 2009 – Spring Update." Directorate-General for Competition.

2015. "Commission Refers Italy to Court for Failure to Recover Illegal State Aid." April 29. http://europa.eu/rapid/press-release_IP-15-4872%5Fen.htm.

Evans, Carolyn L. 2009. "A Protectionist Bias in Majoritarian Politics: An Empirical Investigation." *Economics & Politics* 21 (2): 278–307. doi:10.1111/j.1468-0343.2009.00346.x.

Fan, Shenggen and Neetha Rao. 2003. *Public Spending in Developing Countries: Trends*. EPTD Discussion Paper. No. 99.

Farrell, Sean. 2016. "EU to Investigate Chinese Steel Subsidies Blamed for Dumping." *The Guardian*, May 13, sec. Business. www.theguardian.com/business/2016/may/13/eu-to-investigate-chinese-steel-subsidies-blamed-for-dumping.

Fenno, Richard F. 1978. *Home Style: House Members in Their Districts*. Boston, Mass.: Little, Brown.

Ferejohn, John A. 1974. *Pork Barrel Politics: Rivers and Harbors Legislation, 1947–1968*. Stanford, Calif.: Stanford University Press.

Ferrara, Federico and Erik S. Herron. 2005. "Going It Alone? Strategic Entry under Mixed Electoral Rules." *American Journal of Political Science* 49 (1): 16–31. doi:10.1111/j.0092-5853.2005.00107.x.

Fiorina, Morris P. 1973. "Electoral Margins, Constituency Influence, and Policy Moderation: A Critical Assessment." *American Politics Quarterly* 1 (4): 479–498. doi:10.1177/1532673X7300100403.

Fløysand, Arnt and Stig-Erik Jakobsen. 1999. "The Norwegian Fish Processing Industry: Regional Adaptation and National Policy Implications." http://brage.bibsys.no/xmlui/handle/11250/162302.

Foroohar, Rana. 2017. "Trump Aims for an Industrial Policy that Works for America." *Financial Times*, Monday May 7, 2017. www.ft.com/content/9b6ed79a-318c-11e7-9555-23ef563ecf9a

Ford, Robert and Wim Suyker. 1990. "Industrial Subsidies in the OECD Economies." OECD Economics Department Working Papers. Paris: Organisation for Economic Co-operation and Development. www.oecd-ilibrary.org/content/workingpaper/062357858637.

Fox, Benjamin. 2013. "EU Pleases France, Widens Film Subsidy Rules." November 15. https://euobserver.com/news/122114.

Franchino, Fabio and Marco Mainenti. 2016. "The Electoral Foundations to Noncompliance: Addressing the Puzzle of Unlawful State Aid in the European Union." *Journal of Public Policy*. 36 (3): 407–436.

Frey, Hugo. 2014. *Nationalism and the Cinema in France: Political Mythologies and Film Events, 1945–1995*. New York: Berghahn Books.

Frieden, Jeffry A., David A. Lake, and Kenneth A. Schultz. 2010. *World Politics: Interests, Interactions, Institutions*. New York, N.Y.: W.W. Norton.

Fryde, E. B. 1983. *Studies in Medieval Trade and Finance: History Series*. London: Hambledon Press.

Gallagher, Michael. 1991. "Proportionality, Disproportionality and Electoral Systems." *Electoral Studies* 10 (1): 33–51. doi:10.1016/0261-3794(91)90004-C.

Gallagher, Michael, Michael Laver, and Peter Mair. 2006. *Representative Government in Modern Europe*. New York, N.Y.: McGraw-Hill.

Garrett, Geoffrey. 2001. "Globalization and Government Spending around the World." *Studies in Comparative International Development* 35 (4): 3–29. doi:10.1007/BF02732706.

Ghosh, Jayati. 2014. "India Faces Criticism for Blocking Global Trade Deal, but Is It Justified?" *The Guardian*, August 22, sec. Global development. www.theguardian.com/global-development/poverty-matters/2014/aug/22/india-criticism-blocking-global-trade-deal.

Glaeser, Edward L., ed. 2010. *Agglomeration Economics*. Chicago, Ill.: University of Chicago Press.

Golden, Miriam and Brian Min. 2013. "Distributive Politics Around the World." *Annual Review of Political Science* 16: 73–99. doi:10.1146/annurev-polisci-052209-121553.

Golden, Miriam and Lucio Picci. 2008. "Pork-Barrel Politics in Postwar Italy, 1953–94." *American Journal of Political Science* 52 (2): 268–89. doi:10.1111/j.1540-5907.2007.00312.x.

Golder, Matt. 2005. "Democratic Electoral Systems around the World, 1946–2000." *Electoral Studies* 24 (1): 103–121.

Grier, Kevin B., Michael C. Munger, and Brian E. Roberts. 1994. "The Determinants of Industry Political Activity, 1978–1986." *American Political Science Review* 88 (4): 911–926. doi:10.2307/2082716.

Grossman, Gene M. and Elhanan Helpman. 1994. "Protection for Sale." *American Economic Review* 84 (4): 833–50.

2002. *Interest Groups and Trade Policy*. Princeton, N.J.: Princeton University Press.

2005. "A Protectionist Bias in Majoritarian Politics." *The Quarterly Journal of Economics* 120 (4): 1239–1282.

Hankla, Charles R. 2006. "Party Strength and International Trade A Cross-National Analysis." *Comparative Political Studies* 39 (9): 1133–1356. doi:10.1177/0010414005281936.

Hansen, Wendy L. 1990. "The International Trade Commission and the Politics of Protectionism." *American Political Science Review* 84 (1): 21–46. doi:10.2307/1963628.

Hatfield, John William and William R. Hauk Jr. 2014. "Electoral Regime and Trade Policy." *Journal of Comparative Economics* 42 (3): 518–534. doi:10.1016/j.jce.2014.04.003.

Hauk, William R. 2011. "Protection with Many Sellers: An Application to Legislatures with Malapportionment." *Economics & Politics* 23 (3): 313–344. doi:10.1111/j.1468-0343.2011.00387.x.

Hiscox, Michael J. 2002. *International Trade and Political Conflict: Commerce, Coalitions, and Mobility*. Princeton, N.J.: Princeton University Press.

Horowitz, Donald L. 1985. *Ethnic Groups in Conflict*. Berkeley, Calif.: University of California Press.

Huber, Evelyne, Charles Ragin, and John D. Stephens. 1993. "Social Democracy, Christian Democracy, Constitutional Structure, and the Welfare State." *American Journal of Sociology* 99 (3): 711–749.

Huber, John D. 2000. "Delegation to Civil Servants in Parliamentary Democracies." *European Journal of Political Research* 37 (3): 397–413. doi:10.1023/A:1007033306962.

Huber, John D. and Charles R. Shipan. 2000. "The Costs of Control: Legislators, Agencies, and Transaction Costs." *Legislative Studies Quarterly* 25 (1): 25–52. doi:10.2307/440392.

2002. *Deliberate Discretion?: The Institutional Foundations of Bureaucratic Autonomy*. Cambridge: Cambridge University Press.

Huber, John D., Charles R. Shipan, and Madelaine Pfahler. 2001. "Legislatures and Statutory Control of Bureaucracy." *American Journal of Political Science* 45 (2): 330–345. doi:10.2307/2669344.

Innovation Norway. 2014a. "From Proposal to Profit" http://textlab.io/doc/884608/from-proposal-to-profit.
   2014b. "Annual Report." www.innovasjonnorge.no/contentassets/8bddf4ce20ab4e9797fd1bcaoaca11d1/innovation-norway-annual-report-2014-in-english.pdf.
International Monetary Fund (IMF). 2001. "Government Finance Statistics Manual 2001." www.imf.org/external/pubs/ft/gfs/manual/index.htm.
Inter-Parliamentary Union PARLINE database 2013. http://archive.ipu.org/parline-e/parlinesearch.asp. Last accessed June 13, 2017.
Isaksen, John R. 2000. "Subsidies to the Norwegian Fishing Industry." Norway: Fiskeriforskning. www.nofima.no/filearchive/Rapport%2013-2000%20Subsidies%20-%20an%20update.pdf.
Iversen, Torben and David Soskice. 2006. "Electoral Institutions and the Politics of Coalitions: Why Some Democracies Redistribute More Than Others." *American Political Science Review* (2): 165–181. doi:10.1017/S0003055406062083.
Jensen, Nathen M. 2017. "The Effect of Economic Development Incentives and Clawback Provisions on Job Creation: A Pre-Registered Evaluation of Maryland and Virginia Programs." *Research and Politics* 4 (2).
Johnson, Joel W. and Jessica S. Wallack. 2012. "Electoral Systems and the Personal Vote." hdl:1902.1/17901, Harvard Dataverse, V1.
Johnston, Ron, David Rossiter, and Charles Pattie. 2006. "Disproportionality and Bias in the Results of the 2005 General Election in Great Britain: Evaluating the Electoral System's Impact." *Journal of Elections, Public Opinion and Parties* 16 (1): 37–54. doi:10.1080/13689880500505157.
Jusko, Karen Long. 2017. *Who Speaks for the Poor?* Cambridge: Cambridge University Press.
Kam, Cindy D. and Robert J. Franzese. 2007. *Modeling and Interpreting Interactive Hypotheses in Regression Analysis.* Ann Arbor, Mich.: University of Michigan Press.
Karol, David. 2007. "Does Constituency Size Affect Elected Officials' Trade Policy Preferences?" *Journal of Politics* 69 (2): 483–494. doi:10.1111/j.1468-2508.2007.00545.x.
Katz, Richard S. 1996. "Electoral Reform and the Transformation of Party Politics in Italy." *Party Politics* 2 (1): 31–53.
Katzenstein, Peter J. 1985. *Small States in World Markets: Industrial Policy in Europe.* Ithaca, N.Y.: Cornell University Press.
Kayser, Mark Andreas, and René Lindstädt. 2015. "A Cross-National Measure of Electoral Competitiveness." *Political Analysis* 23 (2): 242–253. doi:10.1093/pan/mpv001.
Kendall, Maurice G. and Alan Stuart. 1950. "The Law of the Cubic Proportion in Election Results." *The British Journal of Sociology* 1 (3): 183–196.
Kienstra, Jeffrey D. 2012. "Cleared for Landing: Airbus, Boeing and the WTO Dispute over Subsidies to Large Civil Aircraft Comment." *Northwestern Journal of International Law & Business* 32: 569–606.
Kies, U., T. Mrosek, and A. Schulte. 2009. "Spatial Analysis of Regional Industrial Clusters in the German Forest Sector." *International Forestry Review* 11 (1): 38–51. doi:10.1505/ifor.11.1.38.

Kolko, Jed. 2010. "Urbanization, Agglomeration, and Coagglomeration of Service Industries," in *Agglomeration Economics*, edited by Edward L. Glaeser. National Bureau of Economic Research Conference Report. Chicago, Ill.: The University of Chicago Press.

Kono, Daniel Yuichi 2006. "Optimal Obfuscation: Democracy and Trade Policy Transparency." *American Political Science Review* 100(3): 369–384. doi:10.1017/S0003055406062241.

2009. "Market Structure, Electoral Institutions, and Trade Policy." *International Studies Quarterly* 53 (4): 885–906. doi:10.1111/j.1468-2478.2009.00561.x.

Kono, Daniel Yuichi and Stephanie J. Rickard. 2014. "Buying National: Democracy, Public Procurement, and International Trade." *International Interactions* 40 (5): 657–682. doi:10.1080/03050629.2014.899220.

Krugman, Paul. 1998. "What's New about the New Economic Geography?" *Oxford Review of Economic Policy* 14 (2): 7–17. doi:10.1093/oxrep/14.2.7.

1990. "Increasing Returns and Economic Geography." Working Paper 3275. National Bureau of Economic Research. www.nber.org/papers/w3275.

1991. *Geography and Trade*. MIT Press. http://search.ebscohost.com/login.aspx?direct=true&scope=site&db=nlebk&db=nlabk&AN=11366.

Lancaster, Thomas D. 1986. "Electoral Structures and Pork Barrel Politics." *International Political Science Review* 7 (1): 67–81. doi:10.1177/019251218600700107.

Lancaster, Thomas D. and W. David Patterson. 1990. "Comparative Pork Barrel Politics: Perceptions from the West German Bundestag." *Comparative Political Studies* 22 (4): 458–477. doi:10.1177/0010414090022004004.

Laver, Michael and Kenneth A Shepsle. 1994. *Cabinet Ministers and Parliamentary Government*. New York, N.Y.: Cambridge University Press.

Leigh, Andrew. 2008. "Bringing Home the Bacon: An Empirical Analysis of the Extent and Effects of Pork-Barreling in Australian Politics." *Public Choice* 137 (1–2): 279–99. doi:10.1007/s11127-008-9327-5.

Levitt, Steven D. and James M. Snyder, Jr. 1997. "The Impact of Federal Spending on House Election Outcomes." *Journal of Political Economy* 105 (1): 30–53. https://doi.org/10.1086/262064.

Library of Economics and Liberty.2016. Accessed July 14, 2017. www.econlib.org/cgi-bin/cite.pl.

Lijphart, Arend. 1999. *Patterns of Democracy: Government Forms and Performance in Thirty-Six Countries*. New Haven, Conn.: Yale University Press.

2004. "Constitutional Design for Divided Societies." *Journal of Democracy* 15 (2): 96–109. doi:10.1353/jod.2004.0029.

Lindbeck, Assar and Jörgen W. Weibull. 1987. "Balanced-Budget Redistribution as the Outcome of Political Competition." *Public Choice* 52 (3): 273–297. doi:10.1007/BF00116710.

Lizzeri, Alessandro, and Nicola Persico. 2001. "The Provision of Public Goods under Alternative Electoral Incentives." *The American Economic Review* 91 (1): 225–239.

Magee, Stephen P., William A. Brock, and Leslie Young. 1989. *Black Hole Tariffs and Endogenous Policy Theory: Political Economy in*

General Equilibrium. Cambridge; New York, N.Y.: Cambridge University Press.

Maloney, William F. and Gaurav Nayyar. 2017. "Industrial Policy, Information, and Government Capacity." Policy Research Working Paper, No. 8056. World Bank, Washington, D.C. World Bank. https://openknowledge.worldbank.org/handle/10986/26735 License: CC BY 3.0 IGO.

Mansfield, Edward D. and Marc L. Busch. 1995. "The Political Economy of Nontariff Barriers: A Cross-National Analysis." *International Organization* 49 (4): 723–749. doi:10.1017/S0020818300028496.

Mansfield, Edward D. and Helen V Milner. 1997. *The Political Economy of Regionalism.* New York, N.Y.: Columbia University Press.

Marshall, Alfred. 1920. *Principles of Economics.* Basingstoke, Hampshire, U.K.: Macmillan and Co.

Martin, Paul S. 2003. "Voting's Rewards: Voter Turnout, Attentive Publics, and Congressional Allocation of Federal Money." *American Journal of Political Science* 47 (1): 110–127. doi:10.1111/1540-5907.00008.

Martin, Shane. 2011a. "Using Parliamentary Questions to Measure Constituency Focus: An Application to the Irish Case." *Political Studies* 59 (2): 472–488. doi:10.1111/j.1467-9248.2011.00885.x.

2011b. "Electoral Institutions, the Personal Vote, and Legislative Organization." *Legislative Studies Quarterly* 36 (3): 339–61. doi:10.1111/j.1939-9162.2011.00018.x.

2014. "Why Electoral Systems Don't Always Matter the Impact of 'mega-Seats' on Legislative Behaviour in Ireland." *Party Politics* 20 (3): 467–479. doi:10.1177/1354068811436061.

Martin, Stephen and Paola Valbonesi. 2008. "Equilibrium State Aid in Integrating Markets." *The B.E. Journal of Economic Analysis & Policy* 8 (1). doi:10.2202/1935-1682.1904.

Mattera, Philip and Kasia Tarczynska. 2015. "Uncle Sam's Favorite Corporations." www.goodjobsfirst.org/sites/default/files/docs/pdf/UncleSamsFavoriteCorporations.pdf.

Matthews, Donald R. and Henry Valen. 1999. *Parliamentary Representation: The Case of the Norwegian Storting.* Columbus, Ohio: Ohio State University Press.

Mayhew, David R. 1974. "Congressional Elections: The Case of the Vanishing Marginals." *Polity* 6 (3): 295–317. doi:10.2307/3233931.

McCubbins, Mathew D., Roger G. Noll, and Barry R. Weingast. 1987. "Administrative Procedures as Instruments of Political Control." *Journal of Law, Economics, & Organization* 3 (2): 243–277.

McCubbins, Mathew D. and Thomas Schwartz. 1984. "Congressional Oversight Overlooked: Police Patrols versus Fire Alarms." *American Journal of Political Science* 28 (1): 165–179. doi:10.2307/2110792.

McGillivray, Fiona. 1997. "Party Discipline as a Determinant of the Endogenous Formation of Tariffs." *American Journal of Political Science* 41 (2): 584–607. doi:10.2307/2111778.

2004. *Privileging Industry: The Comparative Politics of Trade and Industrial Policy.* Princeton, N.J.: Princeton University Press.

Mehiriz, Kaddour, and Richard Marceau. 2013. "The Politics of Intergovernmental Grants in Canada The Case of the Canada-Quebec

Infrastructure Works 2000 Program." *State and Local Government Review* 45 (2): 43–85. doi:10.1177/0160323X13480804.
Meloni, Giulia and Johan Swinnen. 2013. "The Political Economy of European Wine Regulations." *Journal of Wine Economics* 8 (3): 244–284. doi:10.1017/jwe.2013.33.
Midelfart-Knarvik, Karen Helene, and Henry G. Overman. 2002. "Delocation and European Integration: Is Structural Spending Justified?" *Economic Policy* 17 (35): 321–359. doi:10.1111/1468-0327.00091.
Milesi-Ferretti, Gian Maria, Roberto Perotti, and Massimo Rostagno. 2002. "Electoral Systems and Public Spending." *The Quarterly Journal of Economics* 117 (2): 609–657.
Milner, Helen V. 1997. "Industries, Governments and the Creation of Regional Trade Blocs," in *The Political Economy of Regionalism*, edited by Edward D. Mansfield and Helen V. Milner, 77–106. New York, NY: Columbia University Press.
Monroe, Burt L. and Amanda G. Rose. 2002. "Electoral Systems and Unimagined Consequences: Partisan Effects of Districted Proportional Representation." *American Journal of Political Science* 46 (1): 67–89. doi:10.2307/3088415.
Morgenstern, Scott and Stephen M. Swindle. 2005. "Are Politics Local? An Analysis of Voting Patterns in 23 Democracies." *Comparative Political Studies* 38 (2): 143–170. doi:10.1177/0010414004271081.
Mortensen, Dale T. and Christopher Pissarides. 2001. "Taxes, Subsidies and Equilibrium Labor Market Outcomes," in *Designing Inclusion: Tools to Raise Low-end Pay and Employment in Private Enterprise*, edited by Edmund Phelps, 44–73. New York, NY: Cambridge University Press.
Moser, Robert G. 2001. "The Effects of Electoral Systems on Women's Representation in Post-Communist States." *Electoral Studies* 20 (3): 353–369. doi:10.1016/S0261-3794(00)00024-X.
Mozaffar, Shaheen, James R. Scarritt, and Glen Galaich. 2003. "Electoral Institutions, Ethnopolitical Cleavages, and Party Systems in Africa's Emerging Democracies." *American Political Science Review* (3): 379–390. doi:10.1017.S0003055403000753.
Müller, Wolfgang C. 2005. "Austria: A Complex Electoral System with Subtle Effects," in *The Politics of Electoral Systems*, edited by Michael Gallagher and Paul Mitchell, 397–416. Oxford: Oxford University Press.
Naoi, Megumi. 2009. "Shopping for Protection: The Politics of Choosing Trade Instruments in a Partially Legalized World." *International Studies Quarterly* 53 (2): 421–44. doi:10.1111/j.1468-2478.2009.00540.x.
Nielson, Daniel L. 2003. "Supplying Trade Reform: Political Institutions and Liberalization in Middle-Income Presidential Democracies." *American Journal of Political Science* 47 (3): 470–491. doi:10.1111/1540-5907.00034.
Niskanen, William A. 1971. *Bureaucracy and Representative Government*. Chicago, Ill.: Aldine, Atherton.
North, Douglass C. 1990. *Institutions, Institutional Change, and Economic Performance*. Cambridge, N.Y.: Cambridge University Press.
OECD. 1998. *Improving the Environment through Reducing Subsidies*. Paris: Organisation for Economic Co-operation and Development. www.oecd-ilibrary.org/content/book/9789264162679-en.

2007. "Globalisation and Regional Economies: Can OECD Regions Compete in Global Industries? – OECD." www.oecd.org/gov/regional-policy/globalisationandregionaleconomiescanoecdregionscompeteinglobalindustries.htm.

2008. *OECD Regions at a Glance 2007*. Paris: Organisation for Economic Co-operation and Development. www.oecd-ilibrary.org/content/book/reg_glance-2007-en.

2010. "The Interface between Subnational and National Levels of Government," in *Better Regulation in Europe: France*. www.oecd.org/gov/regulatory-policy/45706785.pdf.

OFT. 2007. "Guidance on How to Assess the Competition Effects of Subsidies." Office of Fair Trading. www.gov.uk/government/uploads/system/uploads/attachment_data/file/191490/Green_Book_supplementary_guidance_assessing_compeition_effects_subsidies.pdf.

2009. "Government in Markets Why Competition Matters – a Guide for Policy Makers." Office of Fair Trading. www.gov.uk/government/uploads/system/uploads/attachment_data/file/284451/OFT1113.pdf.

OIV. 2007. "Annual Report." International Organisation of Vine and Wine.

Olson, Mancur. 1965. *The Logic of Collective Action*. Cambridge, Mass.: Harvard University Press.

ONS Nomis Database, Workforce Jobs. www.nomisweb.co.uk/ Last accessed on April 25, 2017.

Owen, Geoffrey. 2012. "Industrial Policy in Europe since the Second World War: What Has Been Learnt?" http://eprints.lse.ac.uk/41902.

Palmer-Rubin, Brian. 2016. "Interest Organizations and Distributive Politics: Small-Business Subsidies in Mexico." *World Development* 84 (August): 97–117. doi:10.1016/j.worlddev.2016.03.019.

Perdikis, Nicholas, Robert Read, and International Economics Study Group. 2005. *The WTO and the Regulation of International Trade: Recent Trade Disputes between the European Union and the United States*. Cheltenham: Edward Elgar Publishing.

Persson, Torsten, Gerard Roland, and Guido Tabellini. 2007. "Electoral Rules and Government Spending in Parliamentary Democracies." *Quarterly Journal of Political Science* 2 (2): 155–88. doi:10.1561/100.00006019.

Persson, Torsten and Guido Tabellini. 1999. "The Size and Scope of Government Comparative Politics with Rational Politicians." *European Economic Review* 43 (4–6): 699–735. doi:10.1016/S0014-2921(98)00131-7.

2003. *The Economic Effects of Constitutions*. Cambridge, Mass.: MIT Press.

Plender, R. 2003. "Definition of Aid," in *The Law of State Aid in the European Union*, edited by A. Biondi, 3–39. Oxford: Oxford University Press.

Porges, Seth. 2013. "6 Things You (Probably) Didn't Know About Cognac." *Forbes*. Accessed August 2, 2016. https://www.forbes.com/sites/sethporges/2013/10/10/6-things-you-probably-didnt-know-about-cognac/#19c8775d2e04.

Powell, G. Bingham Jr. and Georg S. Vanberg. 2000. "Election Laws, Disproportionality and Median Correspondence: Implications for Two Visions of Democracy." *British Journal of Political Science* 30 (3): 383–411.

Pratley, Nick. 2016. "Putting the Industry Back into Departmental Business." *The Guardian*, July 14, sec. Business. www.theguardian.com/business/nils-

pratley-on-finance/2016/jul/14/putting-the-industry-back-into-departmental-business.
Rae, Douglas W. 1967. *The Political Consequences of Electoral Laws*. New Haven, Conn.: Yale University Press.
Rankin, Jennifer. 2016. "EU Steel Action Plan Expected to Include Punitive Tariff Proposal." *The Guardian*, March 15, sec. Business. www.theguardian.com/business/2016/mar/15/eu-steel-action-plan-punitive-tariff-proposal-uk-dumped-goods-subsidised-chinese-imports.
Read, Robert. 2005. "The EU-US WTO Steel Dispute," in *The WTO and Regulation of International Trade*, edited by Nicholas Perdikis and Robert Read, 35–176. Cheltenham: Edward Elgar Publishing.
Reed, Steven R., Ethan Scheiner, and Michael F. Thies. 2012. "The End of LDP Dominance and the Rise of Party-oriented Politics in Japan." *The Journal of Japanese Studies* 38 (2): 353–376.
Remmer, Karen L. 1986. "The Politics of Economic Stabilization: IMF Standby Programs in Latin America, 1954–1984." *Comparative Politics* 19 (1): 1–24. doi:10.2307/421778.
Reynolds, Kara M. and John S. Palatucci. 2012. "Does Trade Adjustment Assistance Make a Difference?" *Contemporary Economic Policy* 30 (1): 43–59. doi:10.1111/j.1465-7287.2010.00247.x.
Rickard, Stephanie J. 2009. "Strategic Targeting The Effect of Institutions and Interests on Distributive Transfers." *Comparative Political Studies* 42 (5): 670–695. doi:10.1177/0010414008328643.
    2010. "Democratic Differences: Electoral Institutions and Compliance with GATT/WTO Agreements." *European Journal of International Relations* 16 (4): 711–729. doi:10.1177/1354066109346890.
    2012a. "Electoral Systems, Voters' Interests and Geographic Dispersion." *British Journal of Political Science* 42 (4): 855–877. doi:10.1017/S0007123412000087.
    2012b. "Welfare versus Subsidies: Governmental Spending Decisions in an Era of Globalization." *The Journal of Politics* 74 (4): 1171–1183. doi:10.1017/S0022381612000680.
    2012c. "A Non-Tariff Protectionist Bias in Majoritarian Politics: Government Subsidies and Electoral Institutions." *International Studies Quarterly* 56 (4): 777–85. doi:10.1111/j.1468-2478.2012.00760.x.
    2015. "Compensating the Losers: An Examination of Congressional Votes on Trade Adjustment Assistance." *International Interactions* 41 (1): 46–60. doi:10.1080/03050629.2015.954697.
Rickard, Stephanie J. and Teri L. Caraway. 2014. "International Negotiations in the Shadow of National Elections." *International Organization* 68 (3): 701–720. doi:10.1017/S0020818314000058.
Rickard, Stephanie J. and Daniel Y. Kono. 2014. "Think Globally, Buy Locally: International Agreements and Government Procurement." *Review of International Organizations* 9 (3): 333–352.
Rodden, Jonathan. 2010. "The Geographic Distribution of Political Preferences." *Annual Review of Political Science* 13 (1): 321–340. doi:10.1146/annurev.polisci.12.031607.092945.

Rodriguez, Francisco, and Dani Rodrik. 2000. *NBER Macroeconomics Annual 2000*. Edited by Ben Bernanke, Kenneth Rogoff, and National Bureau of Economic Research. Cambridge, Mass.: MIT Press.

Rodrik, Dani. 1995. "Political Economy of Trade Policy." *Handbook of International Economics*, edited by Gene Grossman and Kenneth Rogoff, 1457–1494. Elsevier. https://ideas.repec.org/h/eee/intchp/3-28.html.

2004. "Industrial Policy for the Twenty-first Century." Cambridge, MA: Harvard University.

Rogowski, Ronald. 1987. "Trade and the Variety of Democratic Institutions." *International Organization* 41 (2): 203–223. doi:10.1017/S0020818300027442.

Rogowski, Ronald and Mark Andreas Kayser. 2002. "Majoritarian Electoral Systems and Consumer Power: Price-Level Evidence from the OECD Countries." *American Journal of Political Science* 46 (3): 526–539. doi:10.2307/3088397.

Ruddick, Graham. 2016. "Industrial Strategy Welcomed as Part of Government's New Business Policy." *The Guardian*, July 14, 2016. sec. Politics. www.theguardian.com/politics/2016/jul/14/industrial-strategy-welcomed-as-part-of-governments-new-business-policy.

Rudra, Nita. 2002. "Globalization and the Decline of the Welfare State in Less-Developed Countries." *International Organization* 56 (2): 411–445. doi:10.1162/002081802320005522.

Sauger, Nicolas. 2009. "Party Discipline and Coalition Management in the French Parliament." *West European Politics* 32 (2): 310–326. doi:10.1080/01402380802670602.

Schattschneider, E. E. 1935. *Politics, Pressures, and the Tariff*. New York, N.Y.: Arno Press.

Schatz, Klaus-Werner and Frank Wolter. 1987. *Structural Adjustment in the Federal Republic of Germany*. Geneva: International Labour Office.

Schott, Jeffrey J. and Gary Clyde Hufbauer. 2014. "Putting the Trade Facilitation Agreement Back on Track after India's Obstruction." Peterson Institute for International Economics, August 5, 2014. https://piie.com/blogs/trade-investment-policy-watch/putting-trade-facilitation-agreement-back-track-after-indias.

Schrank, William E. 2003. *Introducing Fisheries Subsidies*. FAO Fisheries Technical Paper 437. Rome: Food and Agriculture Organization of the United Nations.

Schwartz, Gerd and Benedict Clements. 1999. "Government Subsidies." *Journal of Economic Surveys* 13 (2): 119–148. doi:10.1111/1467-6419.00079.

Sharp, Margaret. 2003. "Industrial Policy and European Integration: Lessons from Experience in Western Europe over the Last 25 Years." Working/discussion paper. *(Economics Working Papers 30). Centre for the Study of Economic and Social Change in Europe, SSEES, UCL: London, UK*. www.ssees.ucl.ac.uk/wp3osum.htm.

Sharp, Margaret, Geoffrey Shepherd, and David Marsden. 1987. *Managing Change in British Industry*. Geneva: International Labour Office.

Shelburne, Robert C. and Robert W. Bednarzik. 1993. "Geographic Concentration of Trade-Sensitive Employment." *Monthly Labor Review* 116: 3–13.

Shepherd, Geoffrey, François Duchêne, and Christopher Thomas Saunders. 1983. *Europe's Industries: Public and Private Strategies for Change.* Ithaca, N.Y.: Cornell University Press.

Sherman, Natalie. 2017. "Maryland looks to boost business incentives." *The Baltimore Sun.* March 4, 2017. www.baltimoresun.com/business/bs-bz-business-incentive-spending-grows-20170302-story.html.

Shugart, Matthew Søberg. 1999. "Presidentialism, Parliamentarism, and the Provision of Collective Goods in Less-Developed Countries." *Constitutional Political Economy* 10 (1): 53–88. doi:10.1023/A:1009050515209.

Shugart, Matthew Søberg, Melody Ellis Valdini, and Kati Suominen. 2005. "Looking for Locals: Voter Information Demands and Personal Vote-Earning Attributes of Legislators under Proportional Representation." *American Journal of Political Science* 49 (2): 437–449. doi:10.1111/j.0092-5853.2005.00133.x.

Shugart, Matthew Søberg and Martin P Wattenberg. 2001. *Mixed-Member Electoral Systems: The Best of Both Worlds?* New York, N.Y.: Oxford University Press.

Simpson, James. 2011. *Creating Wine: The Emergence of a World Industry, 1840–1914.* Princeton, N.J.: Princeton University Press.

Singer, David Andrew. 2004. "Capital Rules: The Domestic Politics of International Regulatory Harmonization." *International Organization* 58 (3): 531–565. doi:10.1017/S0020818304583042.

Sørensen, Rune J. 2003. "The Political Economy of Intergovernmental Grants: The Norwegian Case." *European Journal of Political Research* 42 (2): 163–195. doi:10.1111/1475-6765.00079.

Spiegel Online International. 2008. "Tought Times in Sweden: Auto Woes Threaten to Wreck Volvo and Saab," December 2.

Statistics Austria. 2000. "Basic Survey on Areas under Vines 1999."

2015a. "Austrian Wine Statistics Report."

2015b. Employment in the Petroleum Industry and Related Industries 2014. Report 2015/48. www.ssb.no/en/arbeid-og-lonn/artikler-og-publikasjoner/_attachment/245120?_ts=150d1fc6aco.

Steenblik, Ronald. 2012. "A Subsidy Primer." Global Subsidies Initiative of the International Institute for Sustainable Development. www.iisd.org/gsi/subsidy-primer/.

2017. A Subsidy Primer. International Institute for Sustainable Development. Global Subsidies Initiative. www.iisd.org/gsi/subsidy-primer. Last accessed June 17, 2017.

Stigler, George. 1971. *The Theory of Economic Regulation.* Chicago, Ill.: University of Chicago.

Stöllinger, Roman and Mario Holzner. 2016. "State Aid and Export Competitiveness in the EU." *Journal of Industry, Competition and Trade*, March, 1–34. doi:10.1007/s10842-016-0222-3.

Stratmann, Thomas and Martin Baur. 2002. "Plurality Rule, Proportional Representation, and the German Bundestag: How Incentives to Pork-Barrel Differ across Electoral Systems." *American Journal of Political Science* 46 (3): 506–14. doi:10.2307/3088395.

Strøm, Kaare. 1992. "Democracy as Political Competition." *The American Behavioral Scientist* 35 (4): 375–396.

2000. "Delegation and Accountability in Parliamentary Democracies." *European Journal of Political Research* 37 (3): 261–290. doi:10.1023/A:1007064803327.

Sturn, Dorothea. 2010. "Decentralized Industrial Policies in Practice: The Case of Austria and Styria." *European Planning Studies* 8 (2): 169–182. doi:10.1080/09654310011081.

Svåsand, Lars, Strøm, Kaare, and Rasch, Bjørn Erik. 1997. "Change and Adaptation in Party Organizations," in *Challenges to Political Parties: The Case of Norway*, edited by Kaare Strøm and Lars Svasand, 91–124. Ann Arbor: University of Michigan Press.

Taagepera, Rein, and Matt Qvortrup. 2011. "Who Gets What, When, How – Through Which Electoral System?" *European Political Science* 11 (2): 244–258. doi:10.1057/eps.2011.35.

Taagepera, Rein and Matthew Søberg Shugart. 1989. *Seats and Votes: The Effects and Determinants of Electoral Systems*. New Haven, Conn.: Yale University Press.

Tavits, Margit. 2009. "Geographically Targeted Spending: Exploring the Electoral Strategies of Incumbent Governments." *European Political Science Review* 1 (1): 103–123. doi:10.1017/S1755773909000034.

Thames, Frank C. and Martin S. Edwards. 2006. "Differentiating Mixed-Member Electoral Systems Mixed-Member Majoritarian and Mixed-Member Proportional Systems and Government Expenditures." *Comparative Political Studies* 39 (7): 905–27. doi:10.1177/0010414005282383.

The Economist. 2008. "Europe's Baleful Bail-Outs." November 1: 62.

"Things to Know about Canada's Dairy Supply Management System | CTV News." 2016. Accessed August 12. www.ctvnews.ca/business/things-to-know-about-canada-s-dairy-supply-management-system-1.2594529.

Thomas, Kenneth P. 2007. *Investment Incentives Growing Use, Uncertain Benefits, Uneven Controls*. Winnipeg, Man.: International Institute for Sustainable Development. www.deslibris.ca/ID/210391.

Thöne, Michael and Stephen Dobroschke. 2008. *WTO Subsidy Notifications: Assessing German Subsidies under the GSI Notification Template Proposed for the WTO*. Winnipeg, Man.: International Institute for Sustainable Development. www.deslibris.ca/ID/213504.c

Tufte, Edward R. 1973. "The Relationship between Seats and Votes in Two-Party Systems." *American Political Science Review* 67 (2): 540–554. doi:10.2307/1958782.

Verdier, Daniel. 1995. "The Politics of Public Aid to Private Industry: The Role of Policy Networks." *Comparative Political Studies* 28 (1): 3–42. doi:10.1177/0010414095028001001.

Vreeland, James Raymond. 2003. *The IMF and Economic Development*. New York, N.Y.: Cambridge University Press.

Wacziarg, Romain and Jessica Seddon Wallack. 2004. "Trade Liberalization and Intersectoral Labor Movements." *Journal of International Economics* 64 (2): 411–39. doi:10.1016/j.jinteco.2003.10.001.

Ward, Andrew. 2009. "Sweden Rules Out State Bail-Out for Saab." *Financial Times*, November 27. www.ft.com/cms/s/0/3aca2404-db7b-11de-9424-00144feabdc0.html#axzz4EsFLCp33.

Weingast, Barry R., Kenneth A. Shepsle, and Christopher Johnsen. 1981. "The Political Economy of Benefits and Costs: A Neoclassical Approach to Distributive Politics." *Journal of Political Economy* 89 (4): 642–664.

Wilson, Rick K. 1986. "An Empirical Test of Preferences for the Political Pork Barrel: District Level Appropriations for River and Harbor Legislation, 1889–1913." *American Journal of Political Science* 30 (4): 729–754. doi:10.2307/2111270.

Winter, Jan A. 1999. "The Rights of Complainants in State Aid Cases: Judicial Review of Commision Decisions Adopted under Article 88 (Ex 93) Ec." *Common Market Law Review* 36 (3): 521–568.

World Trade Organization (WTO) 2006. "Exploring the Links between Subsidies, Trade and the WTO." World Trade Report. Geneva.

Wong, Fayen. 2014. "Steel Industry on Subsidy Life-Support as China Economy Slows." *Reuters UK*. http://uk.reuters.com/article/us-china-economy-steel-idUSKBN0HD2LC20140919.

Yagoda, Maria. 2009. "Cognac in the Crisis." *The Yale Globalist*, May 11, 2009. http://tyglobalist.org/in-the-magazine/theme/cognac-in-the-crisis/.

Zahariadis, Nikolaos. 2001. "Asset Specificity and State Subsidies in Industrialized Countries." *International Studies Quarterly* 45 (4): 603–616. doi:10.1111/0020-8833.00216.

Index

Aaton Company, 47
adjustment costs
  economic geography and, 31–32
  factor mobility and, 56
  geographic concentration and, 93–96
Agreement on Subsidies and Countervailing Measures (WTO) (SCM Agreement), 208–213
agriculture
  geographic concentration in, 29–30
  government partisanship and policies in, 55n.15
  subsidies in, 71n.17
Airbus, 210–213
Anderson, Kim, 71n.17
Appellations d'Origine Contrôlées (AOC) (France), 109–110
Ardelean, Adina, 11–12
asset specificity
  subsidy spending and, 145–156
  geographic concentration and, 93–96
Association Viticole Champenoise, 109–110
audio-visual services industry (France), 46–48, 210–213
Australia
  subsidies in, 1–2, 18–19, 20–21, 43n.4
  wine industry subsidies in, 70–72
Austria
  electoral system in, 120–124
  farm-gate wine subsidy in, 23–24, 120–130, 201–202
  geographic concentration in, 73–74
  non-EU-compliant wine subsidies in, 106
  policy-targeting incentives in, 197–198
  special interests and subsidies in, 33–34
  wine industry subsidies in, 97–99
Austrian People's Party (ÖVP), 127–128
automotive industry, 33–34

ballot access
  open-list systems, 143–144n.8
  measurement of, 79
  geographically concentrated sectors and, 92
Bawn, Kathleen, 96, 145–156
Beauchaud, Jean-Claude, 117–120
Belgium, subsidies in, 170
beneficiaries of subsidies, diversity of, 18
biotechnology industry (United States)
  government subsidies to, 61–62
  economic geography of, 32
Blum-Byrnes agreement (1946), 46–48
Boeing Aircraft, 210–213
Brazil, policy-targeting in, 197–198
British Airways, government subsidies for, 20–21
British Steel, government subsidies for, 20–21
Brülhart, Marius, 27, 30, 75n.24, 76–78
budget process (Norway), 183, 185n.26, 192n.57
bureaucrats
  interviews with, 173
  control of, 184, 186–188
  preferences of, 186
bureaucratic decision-making
  government control over, 184, 186–188
  letters of assignment (Norway), 187
  Norwegian government structure and, 188–190
  Norwegian subsidies and, 184–188
  subsidies and, 173, 185–188
Busch, Marc L., 36–38
business subsidies, 64–65. *See also* industrial subsidies; particularistic economic policies.
Bussereau, Dominique, 111, 114, 116–120

236

Index

Canada
  legislative dynamics in, 56–58
  political effects of geographic
    concentration in, 36–38
  dairy subsidies in, 16–17, 210–213
candidate-centered electoral competition,
    48–51
  Austria and absence of, 124
  ballot systems and, 79
  geographic diffusion and, 92
  legislative dynamics and, 56–58
  list type and district magnitude, 205
  open-list systems, 140–142
capital mobility, economic geography and,
    31–32
Carey, John M., 142–144
Cassing, James, 27–29
causal complexity, 61–62
Center Party (*Senterpartiet*) (Norway),
    188–190
Chang, Eric C. C., 83n.36
"checker board problem," in geographic
    concentration, 74–75
China
  government subsidies in, 20–21
  exports from, 13–15, 34–35
Chirac, Jacques, 117n.38
closed-list systems
  in Austria, 123–124
  characteristics of, 137–138
  cross-electoral district subsidy variations
    in, 25–26
  cross-national subsidy comparisons,
    147–156
  de facto, 142–144
  electoral competitiveness and, 59–60
  economic geography and, 138–140,
    147–156, 169, 202–203
  institutional differences in, 134–137
  measurement of, 142–144
  in Norway, 156–161, 178–179
  personal vote seeking and, 166
  policy outcomes and, 138–140, 170–182
  policy-targeting incentives and, 197–198
  proportional representation systems,
    24–25, 170–182
coalition governments, 156
Cognac producer subsidies, 107–120,
    201–202
  Austrian wine subsidies compared to,
    128–129
  electoral competitiveness and, 117–120
  electoral institutions and, 110–112
  interest group politics and, 109–110
  parliamentary questions about, 113–116
  partisanship and, 112–113, 116–117
concentration, geographic measurement
  cross-national comparisons of subsidy
    spending and, 145–156
  disproportionality and, 88–92
  district magnitude and list type
    measurement, 161–166
  electoral systems, 82–83, 87–88
  manufacturing subsidies, 83–87
  measurements of, 28–29, 144
constitutional design and reform,
    16–17, 207
consumer prices
  impact of subsidies on, 64–69
  in proportional rule democracies, 12–13,
    83n.36
corporate welfare. *See* industrial subsidies.
costs of production, economic geography
    and, 31–32
cotton industry, subsidies for, 39–42
country size
  cross-national comparisons of subsidy
    spending and, 145–156
  manufacturing subsidies and, 88
country-specific geographic concentration
    data, 76–78
  country size and, 81–83
  wine industry subsidies and, 106
Cox, Gary W., 56–58
cross-national comparisons
  economic geography and, 22, 31–32
  electoral competitiveness, 58–60,
    145–156, 171n.2
  electoral systems, 76, 136
  of geographic concentration, 74–75
  government expenditure/subsidy
    spending ratios, 145–156, 201
  government partisanship, 55
  manufacturing subsidies, 72
  methodologies for, 142–156
  of non-EU-compliant subsidies, 99–104
  results analysis, 147–156
  sector-specific subsidies, 70–72, 97–99
  of subsidies, 21–22, 144–156

data availability 73–74
decomposed values
  entropy indices, 75–76
  geographic concentration measurements,
    74–75
de Mesquita, Bueno, 12–13
democratically-elected governments
  politics of economic policies and, 1–3

democratically-elected governments (cont.)
  single member districts and, 79–80
  special interests and, 199–201, 204–206
  subsidy spending in, 97–99
Démocratie libérale (DL) party (France), 116–117
D'Hondt election method, 123, 178–179
dispersion bonus, 9–10, 138–140
disproportionality
  measuring electoral systems and, 79, 88–92
  Norwegian electoral system, 178–179, 203–204
  policy targeting and, 25–26, 203–204
  subsidy budget shares and, 92
distributive policies, 17–21
district magnitude
  economic geography and, 35–36
  empirical expectations concerning, 51–53
  list systems and, 136–137, 161–166
  measurement of, 79–80
  Norwegian district-specific electoral competitiveness and, 190–192
  personal vote seeking and, 162
  in proportional representation systems, 24–25
district-specific electoral competitiveness
  Norwegian subsidy distribution and, 193–194
  Norwegian vote margins and, 190–192
  in plurality-based systems, 46–48, 175
  policy-targeting incentives and, 197–198
  proportional representation systems, 171–173, 175–176
  wine subsidies and, 130–133
Duverger, Maurice, 79–80

e-certificates, as subsidies, (Norway) 66–67
economic geography
  causal complexity and, 61–62
  constitutional design and, 207
  cross-national variations in, 22, 31–32
  defined, 29–30
  electoral district size and, 35–36
  electoral systems and, 3–5, 13–15, 23–24, 39–42, 199–201
  factor mobility and, 56
  government expenditure on subsidies, 169
  government partisanship and, 53–55
  international implications, 208–213
  legislative dynamics and, 56–58
  measurements of, 28–29, 144
  non-economic politics and, 16–17, 206–208
  particularistic economic policies and, 2–3
  plurality electoral systems, 5–7
  politics and, 31–32, 34–35
  proportional representation and, 7–13
  quantitative analysis of, 22
  subsidies and, 17–21
  wine subsidies and, 130–133
  within-country subsidy measurement and, 156–161
economic policy
  democracy and politics of, 1–3, 199–201
  electoral competition and, 39–42, 48–51, 58–60
  electoral institutions and, 39–42
  factor mobility and, 56
  government partisanship and, 53–55
  income and production redistribution and, 27–29
  special interests and, 64–65
  uneven dispersion of employment and production and, 27–29
economic security
  effects of subsidies, 40, 67–70, 130–133
  voters shared preferences and, 3–5
"effective votes," 43–44
electoral competitiveness
  in closed-list PR systems, 170–182, 202–203
  Cognac subsidies and, 117–120
  cross-national comparisons of subsidy spending and, 58–60, 145–156, 171n.2
  defined, 173n.5
  dispersion bonus of subsidies, 50–51
  district-level competitiveness, 58–60, 171–176
  economic policy and, 39–42, 48–51
  empirical expectations concerning, 51–53
  in France, 117–120
  international subsidy restrictions and, 210–213
  legislative dynamics and, 56–58
  measurement of, 188–190
  in Norway, 176–177
  in open-list systems, 197–198, 202–203
  in proportional representation systems, 50–51
  wine subsidies and, 130–133
electoral institutions
  in Austria, 122–124
  comparison of, 134–137
  district size and, 35–36

economic geography and, 3–5, 23–24, 199–201
empirical expectations concerning, 51–53
endogeneity of, 92–93
factor mobility and, 56
in France, 110–112
geographic concentration and, 36–38, 81
government partisanship and, 53–55
institutional differences in, 134–137
legislative dynamics and, 56–58
measurement of, 78–80, 81–83, 88–92
multiple parties in, 96
plurality systems, 5–7
policy outcomes and, 39–42, 204–206
proportional representation and, 44–45
subsidy budget shares and effect of, 88–92
wine subsidies and, 110–112, 120–124, 130–133, 201–202
within-country subsidy measurement and, 156–161
electoral tactics, 196n.66
Ellison, Glenn, 74–75
employment
  economic geography and patterns of, 22, 29–30
  electoral competition and concentration of, 48–51
  geographic concentration of, 31–32, 73–78
  geographic diffusion of, 86
  manufacturing sector, 82
  Norwegian policies for equality in, 190–192
  proportional systems and distribution of, 44–45
  relative concentration benchmarks for, 76
  sector-specific subsidies and, 159
  subsidies' impact on, 40
  uneven distribution of, 27–29
  vote maximization and, 43–44
  wine industry subsidies and levels of, 105–106
entertainment industry, 29–30
entropy indices
  limitations of, 76
  EU employment data, 76–78
  geographic concentration measurement, 75–76
ethnic politics, 16–17, 206–208
European Commission
  bias in rulings by, 101n.9
  member-states subsidy data from, 99–104

non-EU-compliant subsidy cases and, 100–101
reform to subsidy rules by, 213–215
European Council, EU subsidy rules and, 100n.6
European Court of Justice, 100–101
European Union
  Austrian subsidy violation, 122
  case studies of subsidy programs, 99–104
  entropy indices for employment data from, 76–78
  French subsidy violation and, 106
  manufacturing subsidies in, 67–68
  non-EU compliant subsidies cases, 99–104
  State Aid rules in, 23–24, 72, 97–99, 208–215
  subsidies by, 18–19, 66–67, 72
  subsidy wars and, 210–213
  wine industry subsidies and rules of, 106
Evans, Carolyn L., 11–12, 92–93, 151–156
exchange rate manipulation, 68
Export-Import Bank, 210–213
exports, 40

factor mobility, economic geography and, 56
farm-gate wine merchants,
  Austrian subsidies for, 120–130, 201–202
  electoral system and, 122–124
  policy outcome for, 129–130
  politics and, 124–129, 197–198
federalism
  cross-national comparisons of subsidy spending and, 145–156
  electoral system measurement and, 82
Fédération des Syndicats de la Champagne, 109–110
fishing industry, 41, 159, 161
financial sector, 29–30
Finland
  open-list system in, 137–138
  particularistic economic policies in, 1–2
firm-based subsidy targeting (Norway), 184–188
firm location decisions, 32
first-best reelection strategy using subsidies, 51–53
foreign steel imports, duty imposition on, 6
foreign trade
  manufacturing subsidies and reliance on, 87–88
  subsidies and, 19

forest sector
  geographic diffusion of, 10–13, 40–41, 44–45
  subsidies for, 10–13, 40–41, 44–45
France
  Cognac subsidies in, 23–24, 107–120, 170, 201–202
  economic geography in, 29–30
  electoral competition in, 48–51, 117n.38
  electoral institutions in, 110–112
  EU wine subsidies in, 108–109n.18
  film industry in, 46–48, 210–215
  government subsidies in, 20–21
  non-EU-compliant wine subsidies in, 106
  particularistic economic policies in, 1–2
  reform to subsidy policies in, 213–215
  resistance to subsidy restrictions in, 210–213
  wine industry subsidies in, 97–99
Freedom Party of Austria/Freiheitliche Partei Österreichs/FPÖ, 126–127
"Frozen" (film), distribution and marketing in Norway of, 10–13

Gallagher's disproportionality index, 88–92
  Norwegian electoral system and, 178–179
  Least squares index, 79
GDP per capita
  cross-national comparisons of subsidy spending and, 145–156
  electoral system measurement and, 81–83
  subsidies and, 88
geographic concentration
  country-specific measures of, 76–78
  customer base and, 29–30
  causal complexity and, 61–62
  closed-list systems, 169
  defined, 33–34
  district magnitude and list type, 161–166
  electoral competition and, 48–51, 62–63
  electoral systems and, 88–92
  entropy indices, 75–76
  French electoral institutions and, 110–112
  French wine subsidies and, 109–110
  Gini indices, 74–75
  government subsidy spending and, 156–161, 169
  measurement of, 73–78, 158–160
  manufacturing subsidies and, 83–87
  "no-concentration" benchmark for, 76
  in Norway, 158–160
  open-list systems, 147–156, 169
  in plurality systems, 5–7, 23–24, 46–48, 87–92, 147–156, 169
  political consequences of, 204–206
  production factors and, 93–96
  in proportional systems, 36–38, 50–51, 92–93, 169
  relative concentration benchmarks, 76
  special interests and, 169
  trade policies and, 36–38
  of unemployment, 206–208
  vote maximization and, 39–42
  wine industry subsidies and, 97–99, 105–106
  within-country subsidy measurement and, 156–161
geographic diffusion
  closed-list systems, 138–140, 147–156
  district magnitude and list type, 161–166
  electoral competition and, 48–51
  government subsidy spending and, 169, 172–173
  manufacturing subsidies and, 83–87
  in Norway, 160–161
  party-centered vs. candidated-centered systems and, 92
  personal vote seeking and, 162
  production factors and, 93–96
  proportional representation and, 10–13, 23–24, 44–45, 88–92, 147–156, 169, 202–203
  of subsidies, 50–51, 160–161
  vote maximization and, 39–42, 43–44
Germany
  geographically diffuse industries in, 43–44
  geographic diffusion and subsidies in, 44–45
  government subsidies in, 20–21
  mixed-member electoral system in, 78–80
  non-EU-compliant wine subsidies in, 106
  ballot system measurement in, 142–144
  particularistic economic policies and, 1–2
  vote maximization in, 40–41
Gini indexes, geographic concentration measurements and, 74–75
Glaeser, Edward L., 74–75
globalization
  economic geography and, 13–15, 20–21
  politics and, 34–35
  subsidies and, 87–88
  uneven distribution of economic activity and, 27–29

Index

government expenditure on subsidies
  cross-national comparisons,
    145–156, 201
  district magnitude and list type, 161–166
  economic geography and, 199–201
  federal systems and, 82
  geographic concentration and,
    156–161, 169
  geographic diffusion and, 169, 172–173
  list type and district magnitude, 161–166
  Norwegian subsidies, 156–161, 190–192
  plurality-based electoral systems,
    83–87, 169
  proportional representation systems, 22,
    50–51, 83–87, 93–96, 147–156, 169
  ratio of subsidies to total spending,
    67–68, 69–70, 144–156
  reform initiatives for, 213–215
  sector-specific subsidies, 70–72, 183–184
  spending patterns in, 69–70
Grossman, Gene M., 11–12, 27–29,
  179–182

Hankla, Charles R., 35–36
Hansen, Wendy L., 13–15, 36–38
Helpman, Elhanan, 11–12, 27–29, 179–182

ideology, 116–117. *See also* partisanship.
  left-leaning governments, 82–83,
    145–156
imports, 6, 68
income distribution, 27–29
incumbents
  district-level competitiveness and, 175
  legislative dynamics and, 56–58
  in open-list systems, 140–142
  PR *vs.* plurality electoral systems and
    subsidies by, 102–103
  subsidy funding by, 39–42
India
  government subsidies in, 64–65
  trade negotiations and subsidies in, 16–17
industrial policy 20, 54. *See also* sector-
  specific subsidies.
industrial strategy. *See* sector-specific
  subsidies; May, Theresa.
industrial sector
  cross-national subsidy comparisons,
    97–99
  economic and political benefits of
    subsidies in, 20
  electoral district size and, 35–36
  geographic concentration in, 3–5, 13–15
  government partisanship and, 55

stock prices and political influence of,
  69–70
trade policies and geographic
  concentration in, 208
vote maximization and concentration in,
  39–42
Industry Act of 1972 (United Kingdom),
  20–21
inequality
  distribution of subsidies and, 18–19
  uneven distribution of production and
    employment and, 27–29
Innovation Norway, 184–188
international accounting rules, sector-
  specific subsidies and, 70–72
international agreements
  economic geography and, 16–17,
    208–213
  resistance to, 210–213
  restraints on subsidies in, 72, 98–99
  subsidies and, 19
international conflict, 213–215
Inter-Parliamentary Union PARLINE
  database, 5n.9, 143–144n.8
Italy
  EU wine subsidies in, 108–109n.18
  government subsidies in, 20–21
  policy-targeting incentives in, 197–198

Japan,
  subsidies in, 19, 68
  unemployment programs, 208
job security
  economic geography and, 29–30
  subsidies as tool for, 40
Johnson, Joel W., 142–144
Jospin, Lionel, 117n.38, 118–120

Kayser, Mark Andreas, 12–13, 60n.20
Kongeriget Norges Hypotekbank, 176–177
Krugman, Paul, 74–75

labor mobility
  cross-national comparisons of subsidy
    spending and, 145–156
  economic geography and, 31–32
  geographic concentration and, 93–96
Labour Party (DNA) Norway, 188–190
least-squares index, 79
left-leaning governments, 82–83, 145–156
legislative dynamics
  Austrian farm-gate wine merchant
    subsidy, 124–129
  closed-list systems, 138–140

legislative dynamics (cont.)
  economic geography and, 56–58
  French electoral institutions and,
    110–112
  in open-list systems, 140–142
  parliamentary questions, 113–116
  sector-specific subsidies and, 71–72
letters of assignment, for bureaucrats
  (Norway) 187
Liberal Party (Venstre) (Norway), 9–10
Lindstädt, René, 60n.20
list-type, 78–80, 134–137. *See also* closed-
  list systems; open-list systems
  in Austria, 120–123
  district magnitude and, 161–166
  measurement of, 142–144
  policy outcomes and, 138–142
  in proportional representation systems,
    24–25
  variations in, 137–138
lobbying activities
  cross-national comparisons of subsidy
    spending and, 145–156
  French wine industry subsidies and,
    109–110
  government partisanship and, 53–55
  ministry-level decision making and, 56–58
  in Norway, 176n.9
  by trade organizations, 47–48

Maisons de Champagne, 109–110
majoritarian systems, policy-targeting
  incentives and, 11–12. *See also*
  plurality electoral systems.
manufacturing sector
  cross-national comparisons of subsidy
    spending in, 145–156
  employment variations in, 82
  entropy indices for employment in, 75–76
  geographic concentration in, 27–30, 74
  Norwegian subsidies for, 173–176
  subsidies for, 18, 19–20, 64–65, 67–68,
    72, 83–87
  voter turnout and, 195–197
market mechanisms
  cognac producers subsidies and, 112–113
  geographic concentration and, 31–32
Mary Kay (United States), 33–34
May, Theresa, 19–20, 54–55
McGillivray, Fiona, 13–15, 35–38, 171n.3
McKeown, Timothy J., 27–29
mean district magnitude
  measurements of, 79–80
  geographically concentrated sectors and, 92

Meier, Erhard, 125–127
Mexico
  electoral competition and policy in,
    60–61
  government partisanship in, 54
  subsidies in, 71–72
  unemployment program, 208
Milner, Helen V., 13–15, 36–38
minimum threshold requirements, in
  proportional systems, 50–51
mining industry
  electoral competitiveness, 60
  geographic concentration of, 29–30
mixed-member proportional systems
  in Germany, 78–80
  measuring list systems in, 142–144
modifiable areal unit problem (MAUP),
  76–78
Movement for the Defense of Family
  Farmers (MODEF) (France), 112–113
multicolinearity
  cross-national comparisons of subsidy
    spending and, 145–156
  electoral system measurement and, 81–83
multi-party government coalitions
  cross-national comparisons of subsidy
    spending and, 145–156
  subsidy spending in, 96

narrow interests. *See* special interests
National Assembly Elections Act (2000)
  (Slovenia), 40–41
national price levels, 69–70
national vote-share thresholds,
  proportional representation and, 8
nation-wide constituencies, proportional
  representation and, 7
neo-institutional theory, economic
  geography and, 204–206
Netherlands, industry concentration in,
  36–38
New Zealand, vote maximization in,
  40–41
"no-concentration" benchmark
  geographic concentration measurement
  and, 75–76
  relative concentration *vs.*, 76
Nomenclature of Territorial Units for
  Statistics (NUTS), 76–78
nominal-tier elections, 78–80
non-EU-compliant subsidies
  cases involving, 99–104
  predicted number of, 102–103
  proportional representation and, 101–102

Index

non-tariff barriers, 12–13
non-zero tariffs, plurality-based electoral systems, 11–12
North American Free Trade Agreement (NAFTA)
  import protections and, 13–15
  industry concentration and, 36–38
Norway
  bureaucrats in, 173
  construction industry subsidies in, 9, 138–140
  dispersion bonus in, 9–10
  district-level subsidy distribution and vote margins, 193–194
  electoral competitiveness in, 176–177
  electoral systems in, 177–179
  employment concentration in, 29–30
  Farmer's Party in, 55n.15
  firm-based subsidy targeting in, 184–188
  firm location and subsidies in, 32
  geographic concentration measurement in, 76, 152, 158–160
  geographic diffusion of subsidy distribution in, 160–161
  government structure in, 188–190
  income and production redistribution in, 27–29
  lobbying activities in, 176n.9
  manufacturing-sector subsidies in, 190–192
  list type in, 142–144
  policy outcomes in, 170–182
  policy-targeting incentives in, 179–182, 197–198, 203–204
  population density and subsidy distribution in, 195–197
  proportional representation in, 7, 170–182
  sector-specific subsidies in, 71–72, 170–171, 183–184
  subsidies in, 66–67
  variations in subsidies within, 173–176
  vote maximization in, 40–41, 44–45
  within-country subsidy measurement in, 156–161
number of parties in government, 156

Ochs, Jack, 27–29
OECD countries
  agricultural subsidies in, 71n.17
  geographic concentration in, 29–30
Office of Fair Trading (United Kingdom), 40n.2
Oloffson, Maud, 39–42

open-list systems
  ballot access and, 143–144n.8
  characteristics of, 137–138
  cross-national subsidy spending comparisons, 147–156
  empirical expectations concerning, 51–53
  geographic concentration in, 147–156, 169
  geographic diffusion of subsidies in, 202–203
  measurement of, 142–144
  personal vote strategies in, 169
  policy outcomes and, 140–142
  policy-targeting incentives in, 197–198, 202–203
  in proportional representation, 24–25

parliamentary questions (PQs), 113–116
parochial interests. *See* special interests
particularistic economic policies
  absence in Norway of, 176–177
  definitions of, 204–206
  electoral institutions and, 39–42
  plurality electoral systems, 5–7
  proportional representation systems and, 12–13
  subsidies as, 1–3
partisanship. *See also* ideology.
  electoral systems and, 53–55
  French cognac subsidy and, 114
  in Mexico, 54
  policy outcomes and, 53–55
  Socialist Party (France), 114, 117n.38
party-based electoral competition, 48–51
  in Austria, 23, 122–124
  closed-list systems, 138–140, 156–161
  French electoral system and, 48–51, 114, 117n.38, 118–120
  geographic diffusion and, 92
  legislative dynamics and, 56–58
  measurement of, 78–80
  Norwegian electoral system, 178–179
  open-list systems and, 140–142
  vote margins in, 188–190
People's Action Future for Finnmark (*Folkeaksjonen Framtid for Finnmark*), 8, 40–41
People's Republic of China (PRC), government subsidies in, 20–21. *See also* China.
personal vote seeking
  district magnitude and list type, 162–167, 169
  geographic diffusion and, 162

personal vote seeking (cont.)
  list systems and, 140–142, 150
  subsidy spending and, 147–156, 197–198
Persson, Torsten, 11–12, 27–29, 92–93, 151–156
petroleum industry, 76
plurality electoral systems
  conventional wisdom concerning, 11–12
  cross-national comparisons of subsidy spending and, 145–156
  district-level competitiveness in, 46–48, 175
  district size and, 35–36, 80n.33
  economic geography and, 13–15, 32, 36–38
  electoral competitiveness in, 48–51, 58–60, 171–173
  empirical expectations concerning, 51–53
  federalism in, 82
  in France, 110–112
  geographically-concentrated subsidies in, 5–7, 23–24, 46–48, 87–92, 147–156, 169
  government expenditure on subsidies, 83–87, 169
  Grossman and Helpman model, 11–12
  institutional differences in, 134–137
  legislative dynamics and, 56–58
  manufacturing subsidies in, 83–87
  non-EU-compliant subsidies and, 102–103
  party discipline in, 58
  policy outcomes and, 204–206
  reform to subsidy policies in, 213–215
  single-party governments and, 96
  social welfare spending in, 206–208
  special interests in, 64–65
  subsidy budget shares in, 83–87, 169
  vote types in, 43–44
  vote maximization in, 39–42, 46–48
  wine industry subsidies and, 97–99, 105–106, 110–112
Poland, 40–41
polarization in politics, 34–35
policy outcomes
  Austrian farm-gate wine merchant subsidies, 129–130
  closed-list systems, 138–140, 170–182
  district size and, 35–36
  economic geography and, 13–15, 23–24, 31–32, 204–206
  electoral institutions and, 39–42, 204–206
  French Cognac subsidy, 120
  government partisanship and, 53–55
  list-based proportional electoral systems, 138–142
  open-list systems, 140–142
  proportional representation systems, 170–182
  wine subsidies and, 130–133
policy-targeting incentives
  in closed-list systems, 138–140, 156–161
  subsidies and, 25–26
  district magnitude and list type, 161–166
  electoral competition and, 58–60
  empirical expectations concerning, 51–53
  firm-based subsidy targeting, 184–188
  legislative power and, 11
  in Norway, 179–182
  in open-list systems, 140–142
  in plurality-based electoral systems, 11–12
  in proportional representation systems, 171–173, 179–182
  proportional systems and, 197–198
  sector-based targeting, 183–184
  special interests and, 64–65
  subsidies defined as, 66–67
  wine subsidies as, 130–133
politics
  Austrian farm-gate wine merchant subsidy and, 124–129
  Cognac producers subsidies and, 112–113
  economic geography and, 31–32, 34–35
  economic policy and, 39–42
  electoral district size and, 35–36
  EU-compliant subsidies, 100n.6
  manufacturing subsidies and, 72
  of non-EU-compliant subsidies, 101–102
  sector-specific subsidies and, 70–72
  subsidies' effect on, 20, 67–70
  wine industry subsidies, 105–106
population density, subsidy distribution and, 195–197
populism, 15–16
pork, fiscal, 5, 65, 172–176
procurement, by governments, 29, 66
producer subsidies, 1n.2, 64–65
  economic and political benefits of, 64–65
  Norwegian firm-based subsidy targeting and, 184–188
  Norwegian policies for employment equality and, 190–192

Index

production factors
 cross-national comparisons of subsidy spending and, 145–156
 economic geography and costs of, 31–32
 geographic concentration and, 73–78, 93–96
 geographic diffusion and, 93–96
 uneven distribution of, 27–29
"Project Socrates," 54–55
promotional activities, Cognac producer subsidies, 112–113
proportional representation (PR) systems, 3–5
 in Austria, 120–124
 causal complexity, 61–62
 conventional wisdom about, 12–13
 cross-electoral district subsidies and, 25–26
 cross-national comparisons of subsidies in, 24–25, 145–156
 dispersion bonus in, 138–140
 district-level competitiveness in, 171–173, 175–176
 district size and, 35–36, 80n.33, 161–166
 economic geography in, 7–15, 32
 electoral competitiveness in, 50–51, 58–60
 empirical expectations concerning, 51–53
 geographic concentration and, 36–38, 50–51, 92–93, 169
 geographic diffusion and, 10–13, 23–24, 44–45, 88–92, 147–156, 169, 199–201
 government expenditure on subsidies, 22, 50–51, 83–87, 93–96, 147–156, 169
 government partisanship and, 53–55
 institutional differences in, 134–137
 international trade dependency and, 81–83
 left governments associated with, 82–83
 list-type variations, 137–138, 142–144
 measurement of, 78–80
 multi-party governments and, 96
 non-EU compliant subsidies and, 99–104
 in Norway, 7, 170–182
 open lists and subsidy spending in, 147–156
 personal vote seeking in, 162
 policy outcomes in, 170–182, 204–206
 policy-targeted incentives in, 171–173, 179–182, 197–198
 preference votes in, 138
 reform to subsidy policies in, 213–215
 single-country study, 24–25
 social welfare spending in, 206–208
 special interests in, 33–34, 64–65
 subsidy budget share in, 22, 50–51, 83–87, 93–96, 147–156, 169, 202–203
 vote categories in, 43–44
 vote maximization in, 39–42, 44–45
 wine industry subsidies and, 97–99, 105–106
 within-country subsidy measurement and, 156–161

quarrying, geographic concentration of, 29–30
Quentin, Didier, 111, 114

Reagan, Ronald, 54–55
recovery cases, of non-EU-compliant subsidies, 100–101
Reinfeldt, Fredrik, 39–42
Reinhardt, Eric, 36–38
relative concentration measure, of sector-specific employment, 76
rent seeking
 electoral competition and, 58–60
 Norway, 188
 particularistic economic policies and, 2–3
resource redistribution, economic policies and, 1–3
Reynaud, Marie-Line, 111, 114
Roads to Recovery Program (Australia), 43n.4
Rodrik, Dani, 72
Rogowski, Ronald, 12–13, 81–83
Roland, Gerard, 11–12, 27–29
Rosenbluth, Frances, 96, 145–156

Saab (Sweden), 33–34, 39–42
safe districts
 electoral competitiveness and, 58–60
 subsidy distribution and, 193n.58
Sainte Laguë allocation formula, 177–178
Scania AB (Sweden), 33–34
Schreiner, Erich, 126–128
sector employment
 cross-national comparisons of subsidy spending and, 145–156
 economic geography and, 29–30
 manufacturing employment, 82
 Norwegian policies for, 190–192
 subsidy distribution of, 65
sector-specific subsidies
 government priorities for, 70–72
 in Norway, 176n.10
 Norwegian vote margins and, 190–192
 policy-targeting incentives, 183–184
 within-country measurements of, 156–161

service sector, geographic distribution of, 29–30
Shugart, Matthew Soberg, 78–80, 142–144
single member districts (SMDs)
  Cognac producer subsidies politics and, 112–113
  measuring, 79–80
single-party governments
  cross-national comparisons of subsidy spending and, 145–156
  subsidy spending in, 96
single transferable vote systems, 134n.1, 137n.4
Slovenia, 40–41
Social Democrats (SPÖ) party (Austria), 124
Socialist Left Party *(Sosialistisk Venstreparti)* (Norway), 188–190
Socialist Party (France), 114, 117n.38
social welfare programs
  economic geography and, 206–208
  subsidies' impact on spending for, 64–69
Soziale Marktwirtschaft (Germany), 1–2, 20–21
Spain
  EU wine subsidies in, 108–109n.18
  list type in, 142–144
special interests
  democracies' responsiveness to, 199–201, 204–206
  economic geography and, 3–5, 33–34, 204–206
  electoral systems and, 64–65
  French wine subsidies and, 109–110
  geographic concentration and, 82–83, 169
  government subsidies as response to, 69–70
  particularistic economic policies and, 2–3
spending priorities, 64–69
state-owned institutions, 1–2
State Aid. *See* European Union State Aid rules; government expenditure of subsides
statistical analysis of subsidies
  electoral politics and, 22
  list systems and, 24–25
steel industry (United States), 5–7, 33–34
stickiness of government policies, 54n.14
  causal complexity and, 61–62
  subsidy persistence and, 82–83
structural adjustment program, 116
subsidies
  benefits and limitations of, 72
  budgeting process (Norway), 186
  case studies of, selection criteria for, 99–104
  categories of, 66–67
  causal complexity of, 61–62
  in closed-list systems, 139
  cross-district variations in, 25–26
  cross-national comparisons, 20–22, 24–25, 145–156
  dispersion bonus from geographically diffuse subsidies, 9–10
  economic geography and, 17–21
  electoral competitiveness and, 48–51, 58–60, 170–182
  empirical expectations concerning, 51–53
  firm location and, 31–32
  for geographically diffuse industries, 43–44
  statistics on, 67–68
  government partisanship and, 53–55
  international restraints on, 23–24, 72
  in left-leaning governments, 82–83
  measurement of, 144–156
  ministry-level decisions on, 56–58
  Norway's history of, 176–177
  in open-list systems, 140–142
  particularistic economic policies, 1–3
  plurality electoral systems, 5–7, 46–48
  political benefits of, 9, 67–70
  producer's demands for, 1n.2
  in proportional systems, 13, 50–51, 83–87, 93–96
  as share of government expenditure, 67–68
  socioeconomic benefits of, 3, 67–70
  vote maximization and, 39–42
  voter preferences distribution and, 81
  within-country measurements of, 156–161, 173–176
subsidy wars, international agreements and, 210–213
sugar industry (United States)
  subsidies for, 64–69
  vote maximization and concentration in, 39–42
surplus votes
  in Austrian electoral system, 122–124
  economic geography and, 43
  in plural systems, 46–48
  vote maximization and, 43–44
Sweden
  dispersion bonus of subsidies in, 50–51
  district-specific electoral competitiveness in, 191–192n.55
  geographically diffuse groups in, 10–13

open-list system in, 137–138
particularistic economic policies in, 2
special interests and economic geography in, 33–34
subsidies in, 20–21
vote maximization in, 39–42
Swinnen, Johan F. M., 71n.17
Syndicat du Commerce des Vins de Champagne, 109–110

Tabellini, Guido, 11–12, 27–29, 92–93, 151–156
targeted transfers, government spending patterns and, 69–70
tariffs. *See also* trade policy.
  electoral district size and, 35–36
  plurality systems and incentives for, 11–12
  subsidies and, 19, 54n.14, 68, 99
taxpayers, cost of subsidies for, 64–69
tax policy
  Austrian farm-gate wine merchant tax break, 124–129
  Austrian wine subsidies and, 121–122
  government-funded subsidies and, 17–21
  tax incentives in, 66–67
technical support, Cognac producer subsidies, 112–113
Thomas, Kenneth P., 82–83
tourism, Norwegian support for, 10–13
Trade Adjustment Assistance, 206–208
Trade Expansion Act of 1962 (United States), 6
trade organizations, lobbying for subsidies by, 47–48
trade policy
  cross-national comparisons of subsidy spending and, 145–156
  electoral systems and, 81–83
  district size and, 35–36
  factor mobility and, 56, 58
  geographic concentration and, 36–38
  plurality systems and, 11–12, 46–48
  single member districts and, 79–80
  subsidy restrictions and, 19, 210–213
  unemployment and, 206–208
Transatlantic Trade and Investment Partnership (TTIP), 210–213
Trans-Pacific Partnership (TPP), 16–17, 210–213
  subsidies as response to, 19, 68
Transvideo (France), 47
Trump, Donald, 6, 19–20
turnout, voter (Norway) 195–197

unemployment
  economic geography and programs for, 206–208
  Norwegian subsidy programs and, 191n.50
  voting patterns and, 195–197
Union pour la Démocratie Française (UDF) (France), 117n.38, 118–120
United Kingdom
  Chinese imports and Brexit support in, 13–15
  electoral competitiveness in, 60
  film industry in, 213–215
  firm location and subsidies in, 32
  geographic concentration and, 74–75
  government partisanship and policy in, 54–55
  legislative dynamics in, 56–58
  manufacturing subsidies in, 19–20
  particularistic economic policies in, 1–2
  reform to subsidy policies in, 213–215
  sector-specific subsidies in, 70–72
  subsidy programs in, 20–21
  surplus votes and economic geography in, 43
  textile industry subsidies in, 40
  uneven distribution of production and employment in, 27–29
United States
  biotechnology industry in, 32
  district size in, 35–36
  geography of Chinese imports and political outcomes in, 13–15
  government partisanship and policy in, 54–55
  government subsidies and vote maximization in, 39
  import penetration and politics in, 34–35
  legislative dynamics in, 56–58
  manufacturing subsidies in, 19–20
  plurality electoral systems in, 5–7
  political effects of geographic concentration in, 36–38
  special interests and geography in, 33–34
  subsidies in, 18–19
  subsidy wars in, 210–213
  Trade Adjustment Assistance program, 206–208
  uneven distribution of production and employment in, 27–29

Verdier, Daniel, 82–83
Volvo (Sweden), 33–34
vote margins
  electoral competitiveness measurement and, 188–190
  Norwegian manufacturing-sector subsidies and, 190–192
  subsidy distribution in Norway and, 193–194
  turnout and unemployment and, 195–197
vote maximization
  cross-national comparisons of subsidy spending and, 145–156
  economic policy and, 39–42
  electoral competition and, 48–51
  mechanisms for, 43–48
  in open-list systems, 140–142
  in plural systems, 39–42, 46–48
  in proportional systems, 39–42, 44–45
voters' shared preferences
  district magnitude and, 161–166
  economic geography and, 3–5
  geographic distribution of, 81, 204–208
  open-list systems and, 140–142
vote swing
  electoral competition and, 58–60
  proportional representation and, 8–9, 44–45
  vote maximization and, 43–44

wages, subsidies' impact on, 40
Wales (United Kingdom), 29–30
Wallack, Jessica S., 142–144
"wasted votes," 43–44
Wattenberg, Mark P., 78–80
Welfare
  Social, spending in plurality systems, 206–208
  Social, spending and economic geography and, 206–208
  Social, subsidies' impact on spending for, 64–69
wealth redistribution, subsidies and, 17–21
wine industry subsidies
  Austrian farm-gate merchants' subsidy, 120–130, 201–202
  characteristics of, 106–107
  cross-national comparisons of, 97–99
  electoral institutions and, 110–112, 120–124, 201–202
  EU subsidies, 108–109n.18
  exogeneity of geography in, 105–106
  French Cognac producers subsidies, 107–120, 201–202
  politics of, 130–133
Wine Tax Act (Austria), 129–130
winning coalitions, 12–13
within-country measurements
  geographic concentration, 74–75
  policy outcomes, closed-list systems, 170–182
  subsidy spending, 156–161, 173–176
World Trade Organization (WTO)
  Agreement on Subsidies and Countervailing Measures, 208–213
  subsidy regulation by, 23–24, 66–67, 72, 97–99
  trade disputes and, 79–80
  wine subsidies and, 107n.15

Books in the Series (continued from p. ii)

Raymond M. Duch and Randolph T. Stevenson, *The Economic Vote: How Political and Economic Institutions Condition Election Results*
Jean Ensminger, *Making a Market: The Institutional Transformation of an African Society*
David Epstein and Sharyn O'Halloran, *Delegating Powers: A Transaction Cost Politics Approach to Policy Making under Separate Powers*
Kathryn Firmin-Sellers, *The Transformation of Property Rights in the Gold Coast: An Empirical Study Applying Rational Choice Theory*
Clark C. Gibson, *Politicians and Poachers: The Political Economy of Wildlife Policy in Africa*
Daniel W. Gingerich, *Political Institutions and Party-Directed Corruption in South America*
Avner Greif, *Institutions and the Path to the Modern Economy: Lessons from Medieval Trade*
Jeffrey D. Grynaviski, *Partisan Bonds: Political Reputations and Legislative Accountability*
Stephen Haber, Armando Razo, and Noel Maurer, *The Politics of Property Rights: Political Instability, Credible Commitments, and Economic Growth in Mexico, 1876–1929*
Ron Harris, *Industrializing English Law: Entrepreneurship and Business Organization, 1720–1844*
Anna L. Harvey, *Votes without Leverage: Women in American Electoral Politics, 1920–1970*
Murray Horn, *The Political Economy of Public Administration: Institutional Choice in the Public Sector*
John D. Huber, *Rationalizing Parliament: Legislative Institutions and Party Politics in France Jack Knight, Institutions and Social Conflict*
John E. Jackson, Jacek Klich, Krystyna Poznanska, *The Political Economy of Poland's Transition: New Firms and Reform Governments*
Jack Knight, *Institutions and Social Conflict*
Michael Laver and Kenneth Shepsle, eds., *Cabinet Ministers and Parliamentary Government*
Michael Laver and Kenneth Shepsle, eds., *Making and Breaking Governments: Cabinets and Legislatures in Parliamentary Democracies*
Michael Laver and Kenneth Shepsle, eds., *Cabinet Ministers and Parliamentary Government*
Margaret Levi, *Consent, Dissent, and Patriotism*
Brian Levy and Pablo T. Spiller, eds., *Regulations, Institutions, and Commitment: Comparative Studies of Telecommunications*
Leif Lewin, *Ideology and Strategy: A Century of Swedish Politics* (English Edition)
Gary Libecap, *Contracting for Property Rights*
John Londregan, *Legislative Institutions and Ideology in Chile*
Arthur Lupia and Mathew D. McCubbins, *The Democratic Dilemma: Can Citizens Learn What They Need to Know?*
C. Mantzavinos, *Individuals, Institutions, and Markets*
Mathew D. McCubbins and Terry Sullivan, eds., *Congress: Structure and Policy*
Gary J. Miller, *Above Politics: Bureaucratic Discretion and Credible Commitment*

Gary J. Miller, *Managerial Dilemmas: The Political Economy of Hierarchy*
Ilia Murtazashvili, *The Political Economy of the American Frontier*
Douglass C. North, *Institutions, Institutional Change, and Economic Performance*
Elinor Ostrom, *Governing the Commons: The Evolution of Institutions for Collective Action*
Sonal S. Pandya, *Trading Spaces: Foreign Direct Investment Regulation, 1970–2000*
John W. Patty and Elizabeth Maggie Penn, *Social Choice and Legitimacy*
Daniel N. Posner, *Institutions and Ethnic Politics in Africa*
J. Mark Ramseyer, *Odd Markets in Japanese History: Law and Economic Growth*
J. Mark Ramseyer and Frances Rosenbluth, *The Politics of Oligarchy: Institutional Choice in Imperial Japan*
Meredith Rolfe, *Voter Turnout: A Social Theory of Political Participation*
Jean-Laurent Rosenthal, *The Fruits of Revolution: Property Rights, Litigation, and French Agriculture, 1700–1860*
Michael L. Ross, *Timber Booms and Institutional Breakdown in Southeast Asia*
Shanker Satyanath, *Globalization, Politics, and Financial Turmoil: Asia's Banking Crisis*
Norman Schofield, *Architects of Political Change: Constitutional Quandaries and Social Choice Theory*
Norman Schofield and Itai Sened, *Multiparty Democracy: Elections and Legislative Politics*
Alberto Simpser, *Why Governments and Parties Manipulate Elections: Theory, Practice, and Implications*
Alastair Smith, *Election Timing*
Pablo T. Spiller and Mariano Tommasi, *The Instituional Foundations of Public Policy in Argentina: A Transactions Cost Approach*
David Stasavage, *Public Debt and the Birth of the Democratic State: France and Great Britain, 1688–1789*
Charles Stewart III, *Budget Reform Politics: The Design of the Appropriations Process in the House of Representatives, 1865–1921*
George Tsebelis and Jeannette Money, *Bicameralism*
Georg Vanberg, *The Politics of Constitutional Review in Germany*
Nicolas van de Walle, *African Economies and the Politics of Permanent Crisis, 1979–1999*
Stefanie Walter, *Financial Crises and the Politics of Macroeconomic Adjustments*
John Waterbury, *Exposed to Innumerable Delusions: Public Enterprise and State Power in Egypt, India, Mexico, and Turkey*
David L. Weimer, ed., *The Political Economy of Property Rights Institutional Change and Credibility in the Reform of Centrally Planned Economies*

CPSIA information can be obtained
at www.ICGtesting.com
Printed in the USA
LVHW111812150120
643605LV00006BA/742